Food Packaging Handbook

Food Packaging Handbook

Jasmine Harris

MURPHY & MOORE
www.murphy-moorepublishing.com

Food Packaging Handbook
Jasmine Harris
ISBN: 978-1-63987-229-9 (Hardback)

MURPHY & MOORE

Published by Murphy & Moore Publishing,
1 Rockefeller Plaza,
New York City, NY 10020, USA

Cataloging-in-Publication Data

Food packaging handbook / Jasmine Harris.
p. cm.
Includes bibliographical references and index.
ISBN 978-1-63987-229-9
1. Food--Packaging. 2. Active food packaging.
3. Coated foods. I. Harris, Jasmine.
TP374 .F66 2022
664.09--dc23

For more information regarding Murphy & Moore Publishing and its products, please visit the publisher's website www.murphy-moorepublishing.com

Table of Contents

Preface

Food packages are used to enclose food products to protect it from getting damaged, contaminated and spoiled during transportation, storage and sale. These are designed in such a manner that they do not have any adverse impact on food. Labels on food packages are often marked with useful information such as ingredients, nutritional content, amount and cooking instructions. Depending upon the type of container used, they can be classified into primary packaging, secondary packaging and tertiary packaging. Some of the prominent food packaging materials are plastic trays, bags, bottles, cans, cartons and wrappers. Food packaging is an upcoming field of science that has undergone rapid development over the past few decades. Most of the topics introduced herein cover new techniques and the applications of food packaging. This book, with its detailed analyses and data, will prove immensely beneficial to professionals and students involved in this area at various levels.

To facilitate a deeper understanding of the contents of this book a short introduction of every chapter is written below:

Chapter 1- Food packaging is the practice of packing food to protect it from spoilage and to provide it with a special physical and chemical environment. Some of the topics studied under food packaging are functions of food packaging, factors affecting shelf life of food, and packaging design and development. This is an introductory chapter which will briefly introduce all these aspects of food packaging.

Chapter 2- The machinery used in the different packaging operations are known as packaging machinery. Some of the machinery and equipment used for packaging are form fill and seal equipment, cartooning systems, canning equipment, stretch wrap machines, shrink wrapping machines, etc. The topics elaborated in this chapter will help in gaining a better perspective about packaging machinery and equipment.

Chapter 3- There are several types of food packaging technologies such as active and intelligent packaging, aseptic packaging systems, antimicrobial food packaging and modified atmosphere packaging. This chapter discusses in detail the theories and concepts related to food packaging technologies.

Chapter 4- Food packaging materials are the materials that are used to enclose the food to protect it against physical, chemical or biological contamination. The various packaging materials can be classified into metal packaging materials, glass packaging materials, plastic packaging materials, and edible and bio based packaging materials. All these types of packaging materials are described in this chapter in an easy-to-understand manner.

Chapter 5- Packaging of different types of food depend on their individual characteristics and requirements. Packaging of confectionary foods should protect the food from contaminating agents, rancidity, color and aroma loss, and loss of shape. Bags, crates, baskets, cartons and hampers are suitable for packaging of fruits and vegetables. The aim of this chapter is to explore the packaging of various types of food such as beverages, food grains, dairy products, etc.

Finally, I would like to thank the entire team involved in the inception of this book for their valuable time and contribution. This book would not have been possible without their efforts. I would also like to thank my friends and family for their constant support.

Jasmine Harris

1

Food Packaging: An Introduction

Food packaging is the practice of packing food to protect it from spoilage and to provide it with a special physical and chemical environment. Some of the topics studied under food packaging are functions of food packaging, factors affecting shelf life of food, and packaging design and development. This is an introductory chapter which will briefly introduce all these aspects of food packaging.

Food packaging is the enclosing of food to protect it from damage, contamination, spoilage, pest attacks, and tampering, during transport, storage, and retail sale. The package is often labeled with information such as amount of the contents, ingredients, nutritional content, cooking instructions (if relevant), and shelf life. The package needs to be designed and selected in such a manner that there are no adverse interactions between it and the food. Packaging types include bags, bottles, cans, cartons, and trays.

Packaging Types

- Package: It cuts contact between material and outside influences. Package material comes in direct contact with the product.
- Pack: Secondary container. Packing material never comes in contact with product.
- Packing: Number of containers/packages put together in big container is called packing.

Packaging may be looked at as several different types. For example a transport package or distribution package is the package form used to ship, store, and handle the product or inner packages. Some identify a consumer package as one which is directed toward a consumer or household. It is sometimes convenient to categorize packages by layer or function: "primary", secondary", etc.

- Primary packaging is the material that first envelops the product and holds it. This usually is the smallest unit of distribution or use and is the package which is in direct contact with the contents (viz. butter in parchment paper).
- Secondary packaging is outside the primary packaging – perhaps used to group primary packages together (viz. paper board pack containing butter wrapped in veg. parchment paper).
- Tertiary packaging is used for bulk handling, warehouse storage and transport shipping. The most common form is a palletized unit load that packs tightly into containers (viz. Boxes containing 20-25 or 50 butter packs are put together).

Food Packaging Types

The materials mentioned above can be fashioned into different types of food packages and containers. Examples are given below:

Packaging type	Type of container	Examples of foods packaged
Aseptic packages	Primary	Liquid whole eggs
Plastic trays	Primary	Portion of fish.
Bags	Primary	Potato chips.
Bottles	Primary	Bottle of a soft drink.
Boxes	Secondary	Box of soft drink bottles.
Cans	Primary	Can of tomato soup.
Cartons	Primary	Carton of eggs.
Flexible packaging	Primary	Bagged salad.
Pallets	Tertiary	A series of boxes on a single pallet, to transport packaged food from the manufacturing plant to a distribution center.
Wrappers	Tertiary	Used to wrap the boxes on the pallet for transport.

Primary packaging is the main packaging that holds the food that is being processed. Secondary packaging combines the primary packages into a single box. Tertiary packaging combines all of the secondary packages into one pallet.

Special Techniques of Food Packaging

- Vacuum packaging or inert atmosphere packaging: Oxygen in the air tends to reduce the shelf life of food by the process known as oxidation. To prevent this process, some foods are packaged at reduced pressure (partial vacuum) or using an inert gas (such as nitrogen) to replace oxygen.
- Bags-in-boxes: These are used for soft drink syrups, other liquid products, and meat products.
- Wine box: This is a type of box designed for storage of wine.

Packaging Machines

The design and use of packaging machinery needs to take into account the following factors: technical capabilities, labor requirements, worker safety, maintainability, serviceability, reliability, ability to integrate into the packaging line, capital cost, floorspace, flexibility of use, energy usage, quality of outgoing packages, qualifications (for food, phamaceuticals, and so forth), throughput, efficiency, productivity, and ergonomics. Packaging machines may be of the following general types:

- Blister, Skin and Vacuum Packaging Machines.
- Capping, Over-Capping, Lidding, Closing, Seaming and Sealing Machines.

- Cartoning Machines.
- Case and Tray Forming, Packing, Unpacking, Closing and Sealing Machines.
- Check Weighing Machines.
- Cleaning, Sterilizing, Cooling and Drying Machines.
- Conveying, Accumulating and Related Machines.
- Feeding, Orienting, Placing and Related Machines.
- Filling Machines: Handling liquid and powdered products.
- Package Filling and Closing Machines.
- Form, Fill and Seal Machines.
- Inspecting, Detecting and Checkweighing Machines.
- Palletizing, Depalletizing, Pallet Unitizing and Related Machines.
- Product Identification: For labeling, marking and so forth.
- Wrapping Machines.
- Converting Machines.
- Other Specialty Machinery.

Attributes of Packaging

There are also several attributes of packaging that are important:

- Commercially Efficient: One (related to the convenience function) is that it should be efficient from a production or commercial viewpoint, i.e. in filling, closing, handling, transportation and storage.

- Minimal Environmental Impacts: Another is that the package should have, throughout its life cycle from raw material extraction to final disposal after use, minimal negative environmental impacts.

- Not Contaminate Food: A third attribute is that the package should not impart to the food any undesirable contaminants. Although this last attribute may seem self-evident, there has been a long history of so-called food contact substances migrating from the packaging material into the food. For many years the solder used in three-piece metal cans resulted in lead migrating into some foods; the switch to welded side seams eliminated this source of lead. Plasticizers have also migrated into foods, e.g. phthalates, as well as monomers, e.g. styrene monomer, catalysts, e.g. antimony from PET bottles and photoinitiators from printing inks, e.g. ITX. Not surprisingly, food packaging materials are highly regulated to ensure consumer safety.

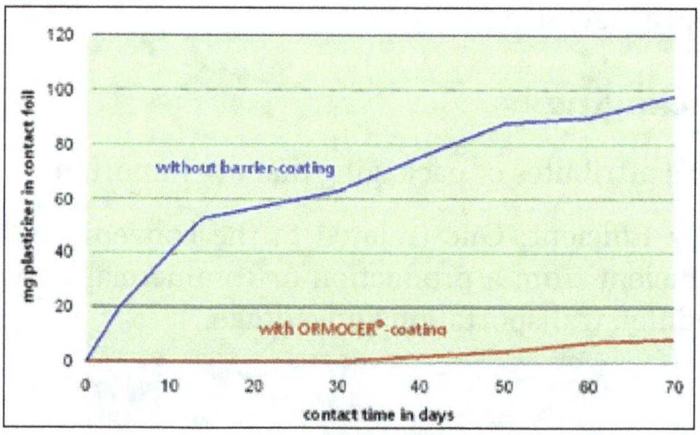

Environments of Packaging

The packaging has to perform its functions in three different environments. Failure to consider all three environments during package development will result in poorly designed packages, increased costs, consumer complaints and even avoidance or rejection of the product by the customer.

Physical

This is the environment in which physical damage can be caused to the product. It includes shocks from drops, falls and bumps; damage from vibrations arising from transportation modes including road, rail, sea and air; and compression and crushing damage arising from stacking during transportation or storage in warehouses, retail outlets and the home environment.

Atmospheric

This is the environment which surrounds the package. Damage to the product can be caused as a result of gases (particularly oxygen), exhaust fumes, water and water vapor, light (particularly UV radiation), and the effects of heat and cold, as well as micro-organisms (bacteria, fungi, molds, yeasts and viruses) and macroorganisms (rodents, insects, mites and birds) which are ubiquitous in many warehouses and retail outlets. Contaminants in the ambient environment such as exhaust fumes from automobiles and dust and dirt can also find their way into the product unless the package acts as an effective barrier.

Human

This is the environment in which the package interacts with people, and designing packages for this environment requires knowledge of the human strengths and frailties of vision, strength, weakness, dexterity, memory, cognitive behavior, etc. It also includes results of human activity such as liability, litigation, legislation and regulation. Since one of the functions of the package is to communicate, it is important that the messages are received clearly by consumers. In addition, the package must contain information required by law such as nutritional content and net weight. To maximize its convenience or utility functions, the package should be simple to hold, open, use and (if appropriate) reclose by the consumer.

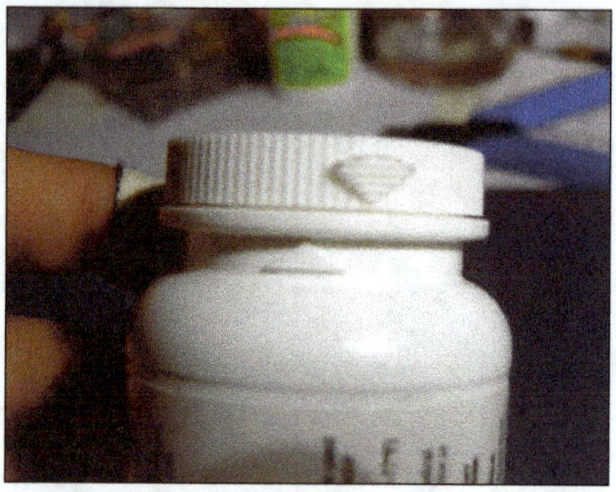

Consumers often complain about the difficulty they experience in opening or accessing packages. Older consumers generally experience more problems than younger consumers, due in large part to the loss of strength as people get older. For example, the grip strength (a measure of the force required undo a cap on a bottle or remove a metal cap on a glass jar) peaks at around 25-39 years for men and women and then declines. The maximum grip strength for men is about 120 pounds, declining to 90 pounds for men aged 60-64 and 75 pounds for men over 70. With an increasing proportion of the population living longer and often alone, packaging manufacturers need to consider this decline in strength with age.

Contributing to the difficulty in opening packages is the trend for more tamper-evident packaging, i.e. a feature on the package which indicates if the package has been tampered with or accessed. The most common form of tamper evidence on a closure is a plastic sleeve which must be breached before the closure can be removed. More sophisticated tamper evident devices use holograms.

Functions/Environments Grid

The functions of packaging and the environments where the package has to perform can be laid out in a two-way matrix or grid as shown in Figure. Everything that is done in packaging can be classified and located in one or more of the 12 function/environment cells. The grid provides a methodical yet simple way of evaluating the suitability of a particular package design before it is actually adopted and put into use. As well, the grid serves as a useful aid when evaluating existing packages including those used by your competitor.

	Physical	Ambient	Human
Containment			
Protection			
Convenience			
Communication			

Food Packaging Materials

The materials used to manufacture food packaging consist of a heterogeneous group including glass, metals, plastics and paper with a corresponding range of performance characteristics. In the selection of suitable packaging materials for a particular food, the focus is typically on the barrier properties of the packaging material. Foods can be classified according to the degree of protection required such as the maximum moisture gain or O_2 uptake. Calculations can then be made to determine whether or not a particular packaging material would provide the necessary barrier required to give the desired product shelf life.

Metal cans and glass containers can be regarded as essentially impermeable to the passage of gases, odors and water vapor provided that a metal end has been correctly seamed on in the case of cans or a satisfactory closure applied in the case of glass containers. Aluminum foil has excellent barrier properties provided that it is at least 25 µm thick; below this thickness the likelihood of pinholes increases. It is common to laminate plastic polymers to aluminum foil to provide mechanical support and heat sealability. Paper-based packaging materials can be regarded as permeable and for this reason are normally coated with a plastic polymer to ensure adequate barrier properties for the packaging of foods. This then leaves plastics-based packaging materials which provide varying degrees of protection, depending largely on the nature of the polymers used in their manufacture.

Food Deterioration

Before suitable packaging can be selected, it is necessary to know how the food deteriorates because in some (but not all) cases, packaging can minimize or slow down the deteriorative reactions. Foods are frequently classified on the basis of their stability as nonperishable, semi-perishable and perishable. An example of the first classification is sugar; provided it is kept dry, at ambient temperature and free from contamination, it should have a very long shelf life. However, few foods are truly nonperishable, and an important factor influencing their perishability is the packaging.

For example, hermetically sealed and heat processed (e.g. canned) foods are generally regarded as nonperishable. However, they may become perishable under certain circumstances, e.g. if the can seams are faulty, or if there is excessive corrosion resulting in internal gas formation and eventual bursting of the can.

Foods with low moisture content such as dried fruits and vegetables, and baked goods are classified as semiperishable. Frozen foods, though basically perishable, may be classified as semiperishable provided that they are properly stored at freezer temperatures.

The majority of foods (e.g. flesh foods such as meat and fish; milk, eggs and most fruits and vegetables) are classified as perishable unless they have been processed in some way. Often, the only form of processing which such foods receive is to be packaged and kept under controlled temperature conditions.

Functions of Food Packaging

The functions of a package are —to preserve the quality and freshness of food, to add appeal to the food to attract consumers, and to facilitate its storage and distribution.|| The basic functions required of a package can be grouped under five major categories.

To Contain the Product

The primary function of any package is to contain the food and facilitate handling, storage, and distribution all the way from the manufacturer to the ultimate user or even the time the rest portion is utilized by the consumer. However, there are usually various levels of packaging. A primary package is one that comes into direct contact with the contained product, e.g., metal cans, glass jars, and plastic pouches. By law, a primary package must not yield any substance that may be injurious to the health of the consumer. Further development to facilitate handling is to bundle a series of primary packages together, and this lead to the concept of secondary packages. Examples of secondary package is corrugated box in which tins of apple juice are packed. As methods of handling and transportation have become more sophisticated, these secondary packages are often palletized and secured by strapping with metal or, more commonly, by shrink- or stretch-wrapped film to give yet another level of packaging, i.e. tertiary packaging. In turn, these pallet loads may be packed into large metal containers, i.e., quaternary packaging for transportation over long distances by air, land, or sea. The secondary, tertiary and quaternary packaging is also known as packing. The following are basic functions during containing:

- Adequate size and shape (trays to support biscuits in package).

- Proper constructional features: No leakage, spillage, diffusion, i.e. loss prevention.
- Package: Must contain the commodity in natural form (chips packed in Pillow pack, prevent damage).
- No subsequent damage after packaging during handling transportation and storage.
- Optimum compatibility (nontoxic, non-soluble with product, No physical, chemical or biochemical changes/alteration, i.e. inert to the product).
- Containment or agglomeration: Small objects are typically grouped together in one package for reasons of efficiency. For example, a single box of 1000 pencils requires less physical handling than 1000 single pencils. Liquids, powders, and granules need containment.

To Protect the Product

One of the most important functions of any container is to protect the product contained against any form of loss, damage, deterioration, spoilage, or contamination that might be encountered throughout the distribution chain. Packaging can prevent physical damage, e.g., bruising caused by vibration shocks during transportation or stacking in a warehouse. Proper packaging will also prevent material loss, e.g., potatoes from a weak sack or juice from a leaky can. Packaging can also protect products against moisture loss or gain, dust, and light, which causes deterioration of some light-sensitive products. It can also protect the package contents against temperature fluctuations in the transit of chilled and frozen foods. Packaging can also be used to control the availability of oxygen to fruits and vegetables and to protect against loss of flavor or fragrance and help products retain their nutritional value.

Proper packaging may also protect the product against microbial spoilage by bacteria, yeasts, and molds. It can also protect against microbiological spoilage of stored products due to rodents and insects. Packaging protects the product against damages which may be due to different hazards viz. (a) Mechanical, (b) Environmental (c) Microbial & Biochemical and (d) Social.

Table: Hazard, damage and protection of packaging materials.

Sr. No	Storage	Hazard	Damage	Protection
I	Handling and transportation	Drop, shunting, shocks, vibrations, stack load, compression etc.	Breakage, loss of shape, dusting, seepage.	Cushioning, blocking.
II	Storage	Stack load, compression, Attack by rodents and insects.	Crushing, distortion sticking, spillage, contamination, spoilage.	Adequate compression strength of package. Resistance ad repulsiveness to insects.

III	Environment during storage transportation and distribution	Biological or otherwise	Contamination	Toughness of packaging material (to resist penetration).
		High/low humidity moisture/water.	Physical, chemical and biological deterioration due to loss/gain of moisture.	Efficiency of closure providing. Water vapour barrier properties. Package desiccant etc.
		O_2	Oxidative rancidity	O_2 Barrier Vacuum – O_2 N_2/CO_2 flushing packaging in impermeable package.
		Light U.V. rays	Vitamin Destruction, Off flavor development, Oxidative rancidity, Bleaching of pigments.	Use of opaque or dark coloured packaging material.
	Storage	Temperature	Change of state, Increase of moisture ingress Increased rate of deterioration.	Heat insulation Use of poor conductor. Use of reflective insulation.
		Time	Gradual and slow changes occur and staling and other deteriorative changes occur.	Early/immediate marketing (FIFO). Proper schedule of dispatching order providing Heat insulation. Use of Barrier material.

Barrier Protection

A barrier from oxygen, water vapor, dust, etc., is often required. Permeation is a critical factor in design. Some packages contain desiccants or Oxygen absorbers to help extend shelf life. Modified atmospheres or controlled atmospheres are also maintained in some food packages. Keeping the contents clean, fresh, and safe for the intended shelf life is a primary function of the package.

Medium of Information

An important function of any food package is to identify the product and its origin; to inform the consumer how to use the contents; to provide any other information needed or required; and very importantly, to attract the user and encourage purchase of the product. Package design has been an important and constantly evolving phenomenon for many years. Marketing communications and graphic design are applied to the surface of the package and in many cases the point of sale/display. The information a package can convey to the consumer may include the following:

- Product manufacturing and best before dates.
- Proper storage conditions.
- Instructions for use.
- Size and number of servings or portions per pack.
- Nutritional information per serving.
- Manufacturer's name and address.
- Cost.
- Suggested recipes.
- Country of origin.
- Information transmission - Packages and labels communicate how to use, transport recycle, or dispose of the package or product.

Means of Minimizing Costs

An important factor often overlooked is that packaging actually reduces costs for the consumer. Packaging reduces food costs by reducing the cost of processing. Foods can be processed where they are grown, waste is treated at the processing plant, and shipping weights are reduced, thereby lowering the cost of transportation. The handling of packages in quantity is important for the economics of bulk storage, warehousing, transport, and distribution. Proper packaging facilitates efficient and mechanized handling, distribution, and marketing of products, thus reducing the high labour costs that would have to be absorbed into the price of the product. Thus, packaging not merely contains the product, but it is a process of bringing goods from the production point to the point of use in a most beneficial manner. This involves all aspects of handling, storage, preservation, distribution, advertising, sales promotion, preparation and various other facts of industry.

Means of Selling Product

The packaging and labels can be used by marketers to encourage potential buyers to purchase the product. Packaging is often referred to as the silent salesman. "A package must protect what it sells and sell what it protects."|| Packages can have features which add convenience in distribution, handling, display, sale, opening, reclosing, use, and reuse. Primary packages should have the following characteristics to facilitate the sale of products:

- Aesthetic appeal.

- Non-toxic.
- Transparent.
- Lightweight.
- Tamper evident.
- Easy to pick up and handle.
- Easy to fit into cupboards, shelves, refrigerators, etc.
- Easy to open and dispense from.
- Easy to reclose.
- Returnable, recyclable, or reusable.
- Safe and presents no hazards in the way of broken glass or sharp jagged metal edges.
- Display the product.
- Glamorize: Create an illusion of something very precious, by decoration, embossing techniques and exotic closures, but it should not deceive the people.

Table: Functional requirements of packaging materials.

No.	Functional Property	Specific Factors
1	Gas permeability	O_2, CO_2, N_2, H_2O vapor.
2	Protection against environmental factors	Light, odor, microorganisms, moisture.
3	Mechanical properties	Weight, elasticity, heat-sealability, mechanical sealability, strength (tensile, tear, impact, bursting).
4	Reactivity with food	Grease, acid, water, color.
5	Marketing-related properties	Attractiveness, printability, cost.
6	Convenience	Disposability, repeated use, resealability, secondary use.
7	Aroma	Aroma compound barrier property.

Functions of a Package

- Dispensing: Product not used all at once, remove a portion, without destroying/damaging the remaining product/container.
- Preserve: Remaining product in container-Protection and preserve it for extended/desired period.
- Measuring/Portion control: Single serving or single dosage package has a precise amount of contents to control usage. Bulk commodities (such as salt) can be

divided into packages that are a more suitable size for individual households. It also aids the control of inventory: selling sealed one-liter-bottles of milk, rather than having people bring their own bottles to fill themselves.

- Security: Packaging can play an important role in reducing the security risks of transport. Packages can be made with improved tamper resistance to deter tampering and also can have tamper-evident features to help indicate tampering. Packages can be engineered to help reduce the risks of package pilferage: Some package constructions are more resistant to pilferage and some have pilfer indicating seals. Packages may include authentication seals to help indicate that the package and contents are not counterfeit. Packages also can include anti-theft devices, such as dye-packs, RFID tags, or electronic article surveillance tags, that can be activated or detected by devices at exit points and require specialized tools to deactivate. Using packaging in this way is a means of loss prevention.

Factors Affecting Shelf Life of Food

Packaging is an essential part of processing and distributing foods. Whereas preservation is the major role of packaging, there are several other functions for packaging, each of which must be understood by the food manufacturer. Packaging must protect against a variety of assaults including microorganisms, insects and rodents. Environmental factors such as oxygen and water vapor will spoil foods if they are allowed to enter packages freely.

Packaging can become a shelf life limiting factor in its own right. For example, this may be as a result of migration of tainting compounds from the packaging into the food or the migration of food components into the packaging. Different groups within the food chain, i.e. consumers, retailers, distributors, manufacturers and growers, proffer subtly different perspectives of shelf life, reflecting the aspect of greatest importance and significance to them. For consumers, it is imperative that products are safe and the quality meets their expectations. Consumers will often actively seek the product on the shelf with the longest remaining shelf life as this is considered to be indicative of freshness.

Shelf Life

The quality of most foods and beverages decreases with storage or holding time. The shelf life of a product is best determined as a part of the product development cycle. The Institute of Food Technologists (IFT) in the United States has defined shelf life as "the period between the manufacture and the retail purchase of a food product, during which time the product is in a state of satisfactory quality in terms of nutritional value, taste, texture and appearance". The Institute of Food Science and Technology (IFST) in the United Kingdom has defined shelf life as "the period of time during which the food product will remain safe; be certain to retain desired sensory, chemical, physical, microbiological and functional characteristics; and comply with any label declaration of nutritional data when stored under the recommended conditions".

The date of minimum durability is defined as the date until which the food retains its specific properties when properly stored. It must be indicated by the words "Best before" followed by the date (or a reference to where the date is given on the labeling). Depending on how long the food can keep, the date can be expressed by the day and the month, the month and the year, or the year alone.

Factors

Product Characteristics

Product characteristics including formulation and processing parameters i.e. intrinsic factors. Intrinsic factors are the properties resulting from the make-up of the final product and include the following:

- Water activity (a_w).
- PH/total acidity.
- Natural micro flora and surviving microbiological counts in final product.
- Availability of oxygen.
- Reduction potential (E_h).
- Natural biochemistry/chemistry of the product.
- Added preservatives (e.g. salt, spices, antioxidants).
- Product formulation.

Environmental Factors

Environment to which the product is exposed during distribution and storage i.e. extrinsic factors. Extrinsic factors are a result of the environment that the product encounters during life and include the following: Temperature, Relative humidity, Gas atmosphere and Light.

Temperature

Temperature is a key factor in determining the rates of deteriorative reactions, and in certain situations the packaging material can affect the temperature of the food. For packages that are stored in refrigerated display cabinets, most of the cooling takes place by conduction and convection. Simultaneously, there is a heat input by radiation from the fluorescent lamps used for lighting. Under these conditions, aluminum foil offers real advantages because of its high reflectivity and high conductivity.

Relative Humidity

The RH of the ambient environment is important and can influence the water activity (a_w) of the food unless the package provides an excellent barrier to water vapor. Many flexible plastic packaging materials provide good moisture barriers, but none is completely impermeable.

Gas Atmosphere

The presence and concentration of gases in the environment surrounding the food have a considerable influence on the growth of microorganisms, and the atmosphere inside the package is often modified. The simplest way of modifying the atmosphere is vacuum packaging, that is, removal of air (and thus O_2) from a package prior to sealing; it can have a beneficial effect by preventing the growth of aerobic microorganisms. Flushing the inside of the package with a gas such as CO_2 or N_2 before sealing is the basis of modified atmosphere packaging (MAP). For example, increased concentrations of gases such as CO_2 are used to retard microbial growth and thus extend the shelf life of foods. MAP is increasing in importance, especially with the packaging of fresh fruits and vegetables, fresh foods, and bakery products.

Atmospheric O_2 generally has a detrimental effect on the nutritive quality of foods, and it is therefore desirable to maintain many types of foods at a low O_2 tension, or at least prevent a continuous supply of O_2 into the package. Lipid oxidation results in the formation of hydroperoxides, peroxides, and epoxides, which will, in turn, oxidize or otherwise react with carotenoids, tocopherols, and ascorbic acid to cause loss of vitamin activity.

With the exception of respiring fruits and vegetables and some fresh foods, changes in the gas atmosphere of packaged foods depend largely on the nature of the package. Adequately sealed metal and glass containers effectively prevent the interchange of gases between the food and the atmosphere. With flexible packaging, however, the diffusion of gases depends not only on the effectiveness of the closure but also on the permeability of the packaging material, which depends primarily on the physicochemical structure of the barrier.

Light

Many deteriorative changes in the nutritional quality of foods are initiated or accelerated by light. Light is, essentially, an electromagnetic vibration in the wavelength range between 4000 and 7000 A, the wavelength of ultraviolet (UV) light ranges between 2000 and 4000 A. The catalytic effects of light are most pronounced in the lower wavelengths of the visible spectrum and in the UV spectrum. The intensity of light and the length of exposure are significant factors in the production of discoloration and flavor defects in packaged foods.

There have been many studies demonstrating the effect of packaging materials with different light-screening properties on the rates of deteriorative reactions in foods. Among the most commonly studied foods has been fluid milk, the extent of off-flavor development being related to the exposure interval, strength of light, and amount of milk surface exposed.

Enzymic Reactions

In food packaging technology, knowledge of enzyme action is essential to a fuller understanding of the implications of different forms of packaging. The importance of enzymes to the food processor is often determined by the conditions prevailing within and outside the food. Control of these conditions is necessary to control enzymic activity during food processing and storage. The major factors useful in controlling enzyme activity are temperature, a_w, pH, chemicals that can inhibit enzyme action, alteration of substrates, alteration of products, and preprocessing control.

Three of these factors are particularly relevant in a packaging context. The first is temperature i.e. the ability of a package to maintain a low product temperature and thus retard enzyme action will often increase product shelf life. The second important factor is a_w, because the rate of enzyme activity is dependent on the amount of water available, low levels of water can severely restrict enzymic activities and even alter the pattern of activity. Finally, alteration of substrate (in particular, the ingress of O_2 into a package) is important in many O_2 dependent reactions that are catalyzed by enzymes, for example, enzymic browning due to oxidation of phenols in fruits and vegetables.

Chemical Reactions

Many of the chemical reactions that occur in foods can lead to deterioration in food quality (both nutritional and sensory) or the impairment of food safety. Such reaction classes can involve different reactants or substrates, depending on the specific food and the particular conditions for processing or storage. The rates of these chemical reactions are dependent on a variety of factors amenable to control by packaging, including light, O_2 concentration, temperature, and a_w. Therefore, the package can, in certain circumstances, play a major role in controlling these factors, and thus indirectly the rate of the deteriorative chemical reactions. The two major chemical changes that occur during the processing and storage of foods and lead to deterioration in sensory quality are lipid oxidation and nonenzymic browning (NEB). Chemical reactions are also responsible for changes in the color and flavor of foods during processing and storage.

- Lipid oxidation: Autoxidation is the reaction of molecular O_2 by a free radical mechanism with hydrocarbons and other compounds. The reaction of free radicals with O_2 is extremely rapid, and many mechanisms for initiation of free radical reactions have been described. The crucial role that autoxidation plays in the development of undesirable flavors and aromas in foods are well documented, and autoxidation is a major cause of food deterioration.

- Nonenzymic browning: Nonenzymic browning (NEB) is one of the major deteriorative chemical reactions that occur during storage of dried and concentrated foods. The NEB or Maillard, reaction can be divided into following three stages:
    - Early maillard reactions involving a simple condensation between an aldehyde (usually a reducing sugar) and an amine (usually a protein or amino acid) without browning.
    - Advanced maillard reactions that lead to the formation of volatile or soluble substances.
    - Final maillard reactions leading to insoluble brown polymers.
- Color changes: Acceptability of color in a given food is influenced by many factors, including cultural, geographical and sociological aspects of the population. However, regardless of these many factors, certain food groups are acceptable only if they fall within a certain color range. The color of many foods is due to the presence of natural pigments such as chlorophylls, anthocyanins, carotenoids, flavonoids, and myoglobin.
- Flavor changes: In fruits and vegetables, enzymically generated compounds derived from long-chain fatty acids play an extremely important role in the formation of characteristic flavors. In addition, these types of reactions can lead to important off-flavors. Enzyme-induced oxidative breakdown of unsaturated fatty acids occurs extensively in plant tissues, and this yields characteristic aromas associated with some ripening fruits and disrupted tissues.

    Aldehydes and ketones are the main volatiles from autoxidation, and these compounds can cause painty, fatty, metallic, papery, and candle like flavors in foods when their concentrations are sufficiently high. However, many of the desirable flavors of cooked and processed foods derive from modest concentrations of these compounds. The permeability of packaging materials is of importance in retaining desirable volatile components within packages and in preventing undesirable components entering the package from the ambient atmosphere.

- Nutritional changes: The four major factors that influence nutrient degradation and can be controlled to varying extents by packaging are light, O_2 concentration, temperature, and aw. However, because of the diverse nature of the various nutrients as well as the chemical heterogeneity within each class of compounds and the complex interactions of these variables, generalizations about nutrient degradation in foods are unhelpful.

Physical Changes

The physical properties of foods can be defined as those properties that lend themselves to description and quantification by physical rather than chemical means and include geometrical, thermal, optical, mechanical, rheological, electrical, and hydrodynamic properties. Geometrical properties encompass the parameters of size, shape, volume, density, and surface area as related to homogeneous food units, as well as geometrical texture characteristics. Although many of these physical properties are important and must be considered in the design and operation of a successful packaging system, in the present context the focus is on undesirable physical changes in packaged foods.

Microbiological Changes

Microorganisms can make both desirable and undesirable changes to the quality of foods, depending on whether they are introduced as an essential part of the food preservation process or arise adventitiously and subsequently grow to produce food spoilage. Every microorganism has a limiting a_w value below which it will not grow, form spores, or produce toxic metabolites. Water activity can influence each of the four main growths cycle phases by its effect on the germination time, the length of the lag phase and the growth rate phase, the size of the stationary population, and the subsequent death rate. Whether a microorganism survives or dies in a low aw environment is influenced by intrinsic factors that are also responsible for its growth at higher a_w. These factors include water-binding properties, nutritive potential, pH, E_h, and the presence of antimicrobial compounds. Microbial growth and survival are not entirely ascribed to reduce aw but are also attributable to the nature of the solute. Key extrinsic factors relating to aw that influence microbial deterioration in foods include temperature, O_2, and chemical treatments. These factors can combine in a complex way to encourage or discourage microbial growth.

2

Packaging Machinery and Equipment

The machinery used in the different packaging operations are known as packaging machinery. Some of the machinery and equipment used for packaging are form fill and seal equipment, cartooning systems, canning equipment, stretch wrap machines, shrink wrapping machines, etc. The topics elaborated in this chapter will help in gaining a better perspective about packaging machinery and equipment.

Form Fill and Seal Equipment

These machines use a reel; of flexible material (paper, film or laminates of paper/film/foil) and either form it into a tube and then seal and fill it at regular intervals, or fold it lengthwise and seal it at right angles to the fold to form a series of pockets which are filled and closed. Form fill and seal machinery may be conveniently divided into three types:

- Vertical machines, in which the material is formed into a circular section tu8be over a forming collar.
- Horizontal machines, in which the material is formed into a rectangular section tube through a forming box.
- Sachet forming machines, in which after filling, the remaining side is sealed to form the complete pack.

The seals in sachet are always are always made between different areas of the same face of the web, but with pouches either face to face or overlap sealing is possible on the long face seal produced in forming the tube.

Vertical form, Fill and Seal Machine

Vertical Form, Fill and Seal (VFFS) machines are automated assembly line packaging systems used in packaging liquids and solids. VFFS machine can be intermittent or continuous motion. Intermittent motion machines operate on the principle that vertical bag seals are made when the film is moving and horizontal seals occur when the film stops.

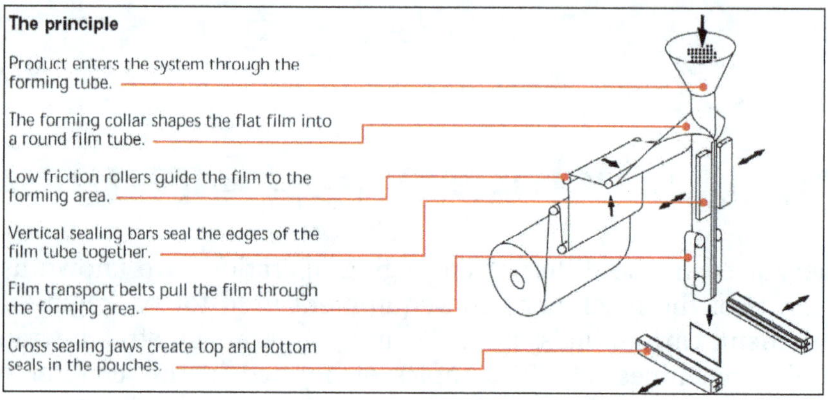

Process

- A roll of film is unwound and formed into a tube over a forming collar. A vertical overlap seal is applied to the tube by the vertical sealing bars. A bottom seal is created by the horizontal sealing bars in the cross jaws.
- The formed pouch is then filled with correct amount of product while the film tube is being fed by the film transport belts. Often a precision scale or an auger ensures that the proper amount of product is placed in the pouch in a consistent manner.
- After the correct length of film tube has been fed, a top seal of the filled pouch is created by the horizontal sealing bars, while the filled pouch is cut from the descending film tube by a knife in the cross jaws.
- If the film is pre-printed, a registration sensor is added to the system to correct the film position in order to maintain the correct print position relative to the end of the pouch. The bottom seal creation of the empty pouch, top seal creation of the filled pouch and cutting of the filled pouch all occur at the same time.

Vertical form fill and seal machine can produce three styles of pack:

- Pillow packs: These characteristically have fin or overlap seals on the base of the pack and transverse seals at either end. The most common applications are solid performed items or multipacks e.g., candy bars or chocolate enrobed biscuits and collections of smaller solid performed items.
- Sachet packs: These have a four sided fin seal around the edge of the pack. The most common applications are [powdered and granular or similar products e.g., instant soups, instant potato, instant desserts.
- Strip packs: This consists of two layers of material sealed together to contain the product between them in individual pockets. The most common applications are in pharmaceuticals: pills, capsules, suppositories.

Vertical form, fill and seal machinery:

- The product enters the system through the forming tube.
- The forming collar shapes the flat film into a round film tube.
- Low friction rollers guide the film to the forming area.
- Vertical sealing bars seal the edges of the film tube together.
- Film transport belts pull the film through the forming area.
- Cross sealing jaws create top and bottom seals in the pouches, and cut the filled pouch from the empty pouch.

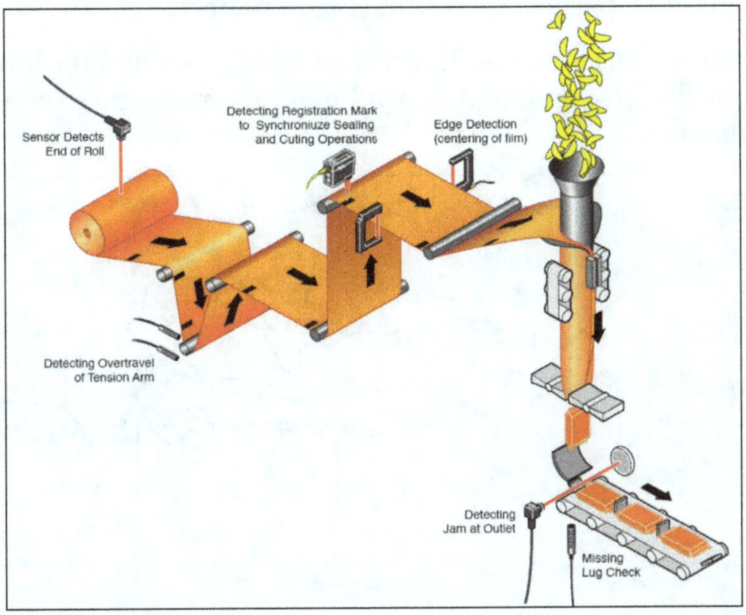

Applications of VFFS

Vertical Form, Fill, and Seal (VFFS) machines are used in the consumer products industry for a wide variety of packaging applications. Various products like salt, tea, sugar, spices, snack foods, wafers, detergent and candies are placed into formed pouches and then sealed. The pouch material is flexible and typically heat-sealable plastic. Paper is also used and sealed by glue.

Horizontal form Fill and Seal Machine

These have much in common with their vertical counterparts. The horizontal machine combines the three separate operations of pack forming or shaping, product introduction into the pack and pack closure. The process also uses a reel or reels of wrapping material; the principal difference is that material is pulled from the reel into a horizontal plane where the subsequent operations take place.

Process

- Sequence of operations of a typical basic machine: The products are loaded into a continuously moving infeed conveyor. The type of conveyor is dependent on the product, but a chain-lug infeed is the most commonly used.
- The wrapping material is formed into a tube inside a folding box. The edges of tube are fin sealed together as they pass between a pair of horizontally mounted propelling and sealing rollers.
- Automatic product feeding: For pouch pack machines these may range from relatively simple units for the more common applications, to custom-designed systems to suit a particular individual requirement.
- Alternative sealing systems: The type of folding box and the layout of the long seam propelling/sealing unit depend upon the packaging application and the materials used.

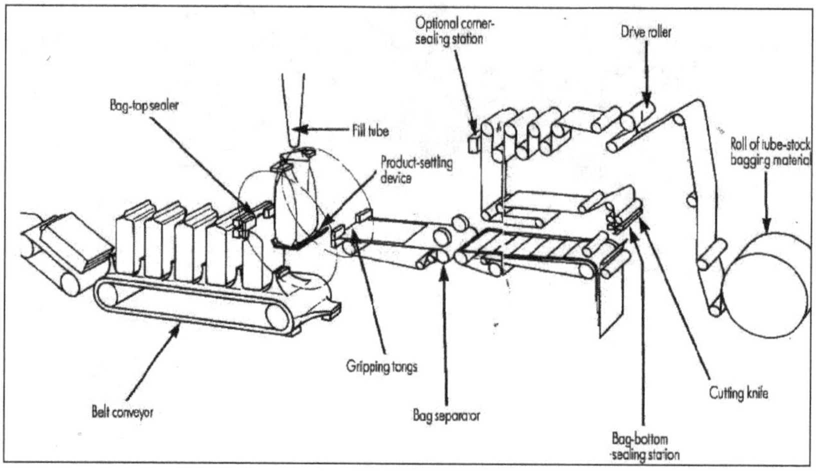

Advantages of FFS

- Bags made on FFS are less expensive than pre-made bags = "pay back."
- Lower capital cost than fully auto machine when packing rate above 500 bags/h.
- Total automation - usually unattended, labour saving provides = "pay-back."
- Consistent packing rate.
- Compact (especially the Vertical FFS).
- Fast - increased production rate, reduced no. of shifts = "pay back."
- Easy and quick to clean-out between production batches.
- Fewer moving parts - especially Vertical FFS.
- 'Online' film printing provides readable barcode and product info.

Disadvantages of FFS

- Not economical for short production runs of one bag size - below 1000 x 25 Kg bags = 25 tons.
- Cannot handle paper.
- Cannot handle multi-ply packing materials.
- Not economical for packing low density, aerated powders such as wheat flour, talcum powder, skimmed and whey powder at medium speeds.
- Cannot remove 'trapped air' very quickly or easily.
- Shape of package made by Vertical FFS or 'C' type FFS not as good as valve sack.
- Cannot self-adjust for different widths of bag.

Cartoning System

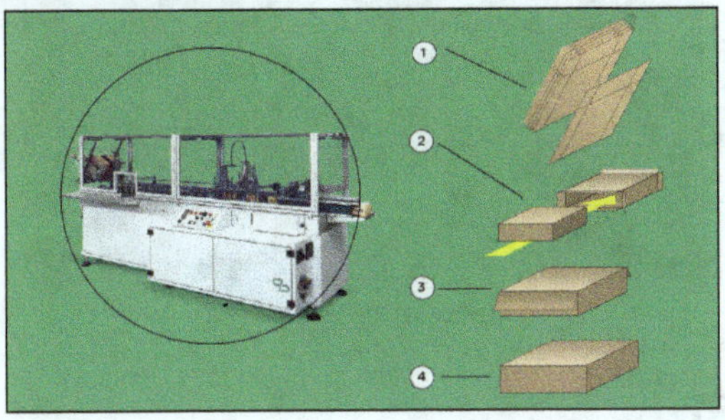

A cartoning machine is an integration of air, electrical, light and mechanical high-tech equipment configured to a computer to control, manage and regulate its operation. The electrical energy is the most essential component of the cartoning machine operation. It is because it powers all the electrical circuit which set in motion all other systems.

For instance, movements of different parts are specifically designed to produce certain desired functions such as circular motion at particular instances form the mechanical part. It is important to understand better and know everything about Cartoning machines. Fundamentally, it works by forming cartons:

- It shapes cartons out of blanks to stand up-straight.
- Close or/and erect.
- Introduce a product.
- Folds.
- Side-seams and finish by sealing.

A Step-by-Step Carton Forming Process

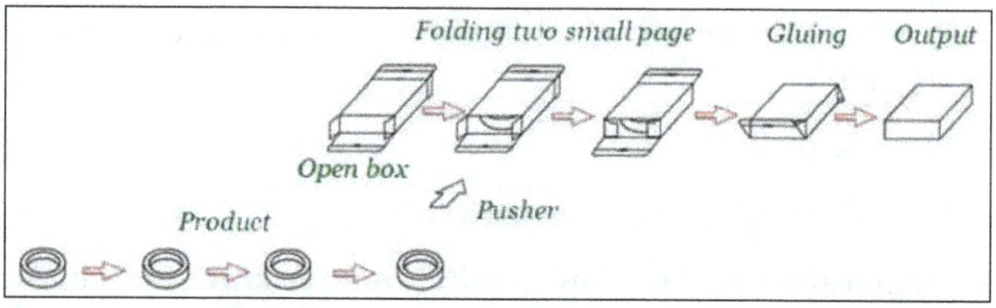

What is evident about a Cartoner that makes it different from other packaging machines? It packages inventory into cartons. For example, have you ever been to supermarket stores? You'll realize that cosmetics or pharmaceutical products are in cartons of different sizes and shapes. Such product packaging is probably done by a Cartoning machine.

Cartoning process.

Carton Shape and Size

Cartoning machine is simply fed with blank stacks of cartons that are shaped to pack inventory. So, what are those cartons made up of and where do they come from? Cartons are made and designed from carton-board: a paper material with semi-flexibility and considerable thickness, used for holding goods especially food or liquids while on transit.

Arrow points at the carton blanks.

Six-sided rectangular or square shaped cartons are the commonly known cartons, but there are still cartons of various shape and size. These wide range of shapes includes double-wall, octagonal, frame-wall, hinged lids, triangular, tapered trays and wave-shaped.

Packaging machine.

You must bear in mind that the method deployed in filling and closing determines what shape and size of a carton to be used. Besides methods of filling and closing the cartons, another significant issue is to get carton to stand up-straight to provide clear space for product insertion and sealing it.

Cartons are supplied with different designs to the buyers and some come with pre-cut slots and tabs; others have the folds glued with adhesive. Mostly, cartons are brought in the form of blanks that are flat, cut to shape and size, pre-printed, and others with tabs and slots pre-cut if requested.

Basically, a Cartoning machine is initially fed or rather, supplied with stacks of fold cartons, leaflet or brochures, or production line. During its operation, the Cartoner picks up a piece, shapes it (or gets it to stand up-straight). It continues to introduce product(s) horizontally or vertically through one open end. Thereafter, it closes by pushing, folding or turning the flaps of the carton and before applying glue or any other suitable adhesive.

Most Cartoning machines use mechanical sleeve or pressurized air to push products into the cartons. But you should not be surprised when you come across Cartoners that require human input to insert products into the carton. The above sequence is performed repeatedly forming a chain of packed cartons moving along the guide rail on a conveyor running to carton chute.

Why you Need a Cartoning Equipment

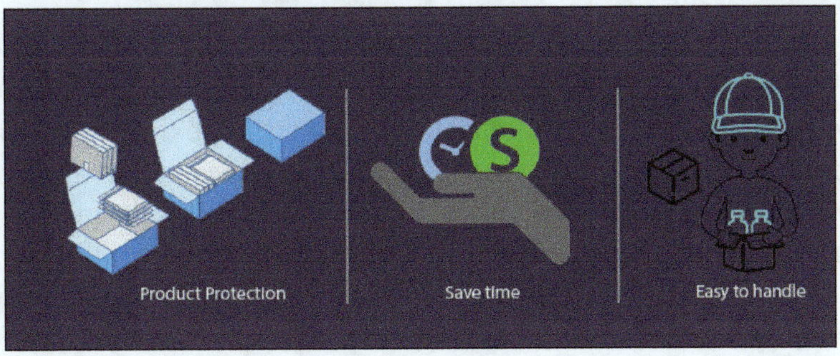

Cartoning equipment have revolutionized how industries package their finished products all over the world. With these machines, they've been able to keep up with ever rising levels of demand for goods such as pharmaceutical products, food stuff and many more. As a result, they've been able to maintain steady production process. This is essential to succeeding in the business; ensuring readily availability of the product for the market.

A number of factors contribute to getting product packaging right. For example, safety procedures, adequate operational training programs, equipment upgrade and preventive maintenance. Therefore, a decisive role has to be played to positively impact on the output of your business.

There are innumerable advantages that come with adaption and implementation of cartoning machine for your business. But if you cannot envisage such importance guaranteed by cartoning equipment; then below are a few reasons to why you need a cartoning equipment in a production environment.

Product Protection and Ease of Transport

Carton provides protection to products filled in them. Since cartons assume regular shapes, makes it easier for bulking packaging for lorries, containers and other vessels easing transportation. For instance, bottles need a proper carriage to accommodate its unique shape. However, with an appropriate cartoning machine such as ZH-100P Cartoning Machine for Bottle&Vial we can achieve a better way to pack bottles.

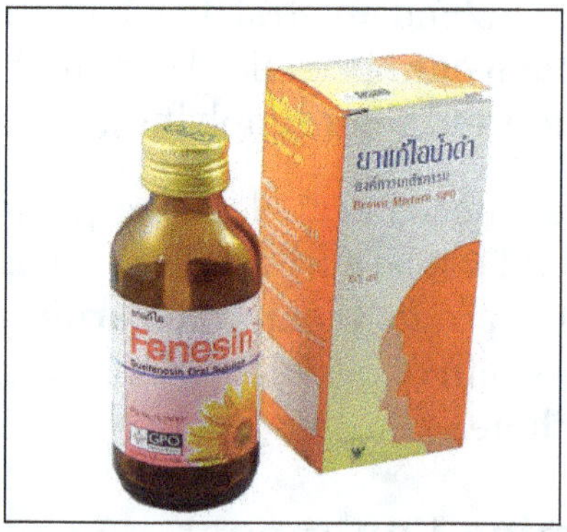

It is easier to package and transport bottles in boxes.

Save Time

High speed production lines are susceptible to time wastage during packaging. However, with continuous mode cartoning machine high speed is achievable and guarantees high throughput altogether. In case you have gone for a fully-automatic cartoning machine that means there will be reduced labor-input during the production process. And since machine does not suffer fatigue unlike man, production can run for longer hours without unnecessary breaks, ultimately, high throughput.

Automatic cartoning machine.

Tamper Proof

Apart from protecting the products from adulteration and interference, carton packaging can also act as tamper proof. Carton packaging done by aid of human and a machine show stark differences from ones done with a Cartoner. A cartoning equipment are designed to perform each task in trig and typically same manner, which can be vividly observed. The unique manner in which the cartoning equipment seal the cartons can aid in determining whether there is tampering with any product. This reason also builds consumer confidence in your brand.

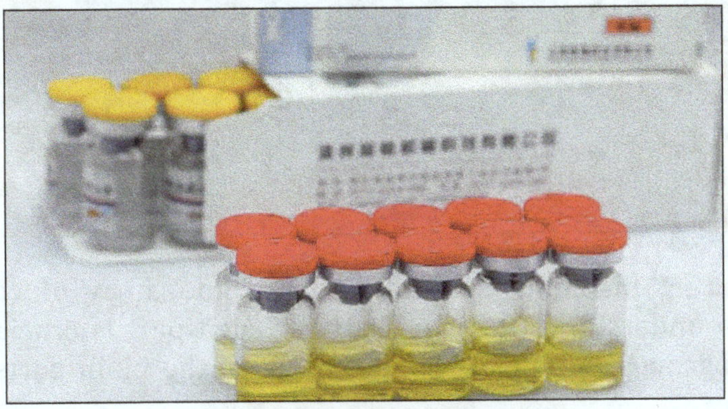

Packaging ensures safety.

Handle a Vast Array of Commodities

If you are concerned about multi-packaging, then you should cast away your aspersions as some semi-automatic cartoning equipment can multitask. For instance, products such as tooth paste and brush can be simultaneously packaged into single carton.

Besides, bunch packaging can also be carried out effectively and efficiently. Thus, investment on cartoning equipment proves to be a real game changer when planning to swiftly move your inventory to market. Since you have understood why you need a cartoning equipment for your business, let's continue to explore more about cartoning machines.

Packaging different products.

Main Parts of a Cartoning Machine

Cartoning machine is an assembly of devises comprising of electrical, mechanical, pneumatic as well as light components. They are all combined to work together in such a manner so as to produce, transform or rather utilize effectively mechanical energy generated from the system.

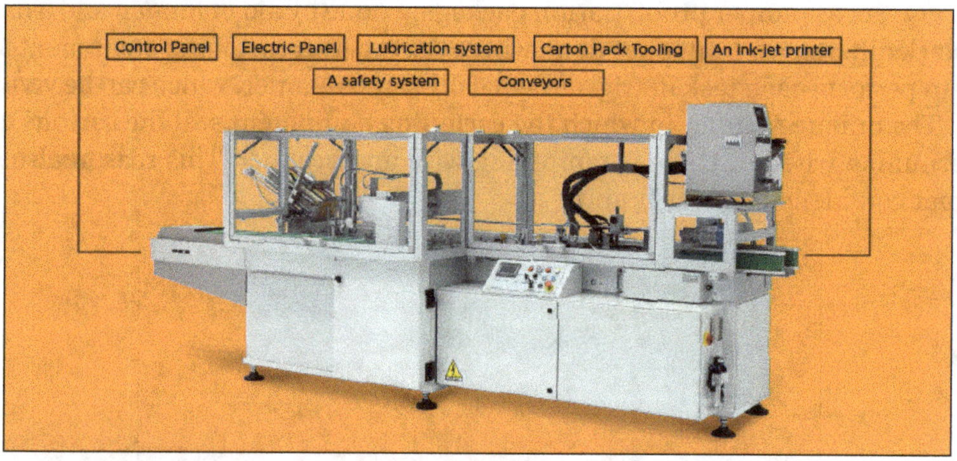

Indeed, a cartoning machine is a setup of a wide array of components and parts. Getting to understand the nitty-gritty of each part is crucial in choosing a suitable cartoning machine. Below is an outline of the main parts of a cartoning machine:

- Control Panel/segment.
- Electric Motors & Electric Panel.
- Lubrication system.
- Carton Pack Tooling & Pneumatic System.
- An ink-jet printer.
- A safety system.
- Conveyors, Rollers, Guide Rail and Carton Chute.

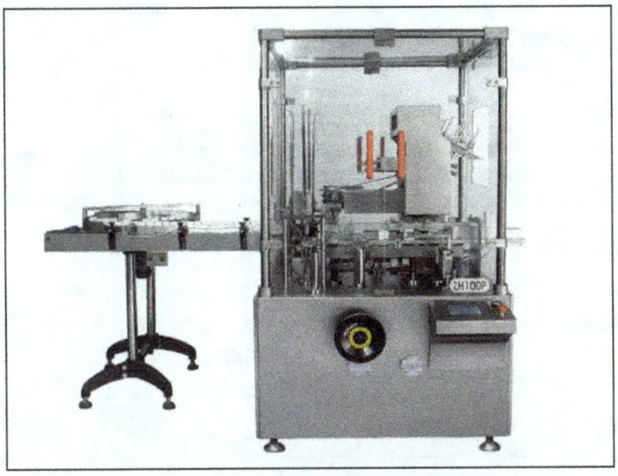

Cartoning Machine for Bottle & Vial.

Control Panel/Segment

With the advent of new technology, integration of various systems has been realized reducing cost involved in assigning individuals to monitor such processes. This forms one of the very essential components of a cartoning machine. The whole process has been configured and design to a specific automation and control module (PLC). A more sophisticated Cartoner control segment can perform and monitor the following:

- Number of cartons per control.
- Number of products per-pack control.
- Print control.
- Electrical and fluid indicators.
- Product Placement Control.

Find below a visual image of the Human Control Interface and a display of its menu:

Touchscreen control panel.

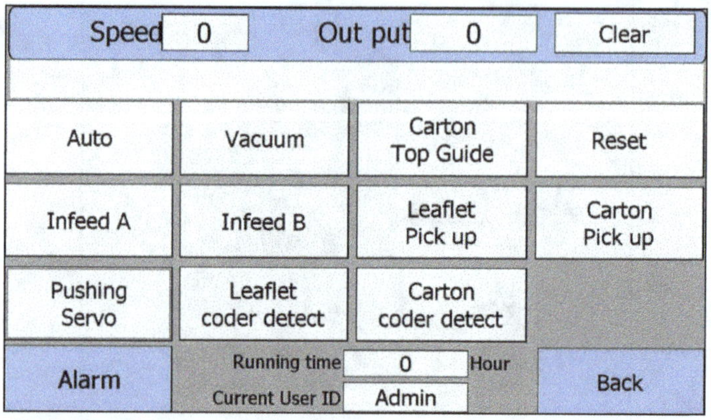

The display on the touchscreen control panel for the cartoning machine.

Is it possible to control the whole cartoning process from the panel? Yes, it is possible to manage and regulate production process from the control segment. In layman's term the control panel is the 'brain' of the machine. Siemens have made it easier by integrating the frequency-converter, PLC (Programmable Logic Controller) with an interface called Human Machine Interface (HMI).

The Human Machine Interface enables an operator to instruct, revoke even intuitively stop all operations if deemed necessary. The HMI also allows the operator to change parameters and requirements of production process in order to achieve a desired set point within certain duration. Beside the controlling aspect, the panel is also configured to alert the operator in case of breakdowns from other parts of the machine.

Electric Panel and Electric Motors

The mains for the cartoning machine are tapped at the electric panel. Electric system also powers up the pneumatic system to function. Always remember that the electric panel carries all the connections to the whole cartoning machine. This section is the 'heart' of the machine since all electrical circuit and mechanical system are configured here.

It is important to always have an electrical schematic diagram for a cartoning machine to assist wherever necessary in the event of an electrical fault. The image below is an electrical panel where all circuits are configured.

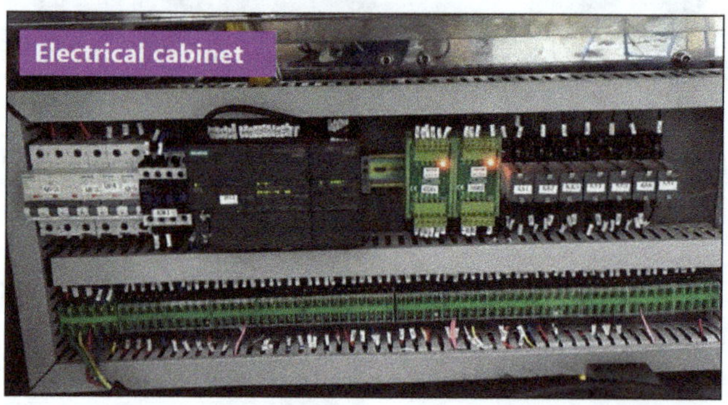

Electric Motor

It is a device that converts electric energy into mechanical energy. An electric motor is a device that uses electricity to generate a desired movement. All motions and movements produced by a cartoning machine when running is solely attributed to functioning of motors.

Research and development on motors have opened up many applications that are instrumental in creating different function in the cartoning machine. Like many other machines, motors in a Cartoner are also well enclosed and placed at the bottom section. The chains, bearing, belt, gears and shafts are conveniently connected via gearing axis, support axis to the machine middle shafts to transmit motion to desired sections. Below is a simple schematic diagram of the mechanical system of a cartoning machine:

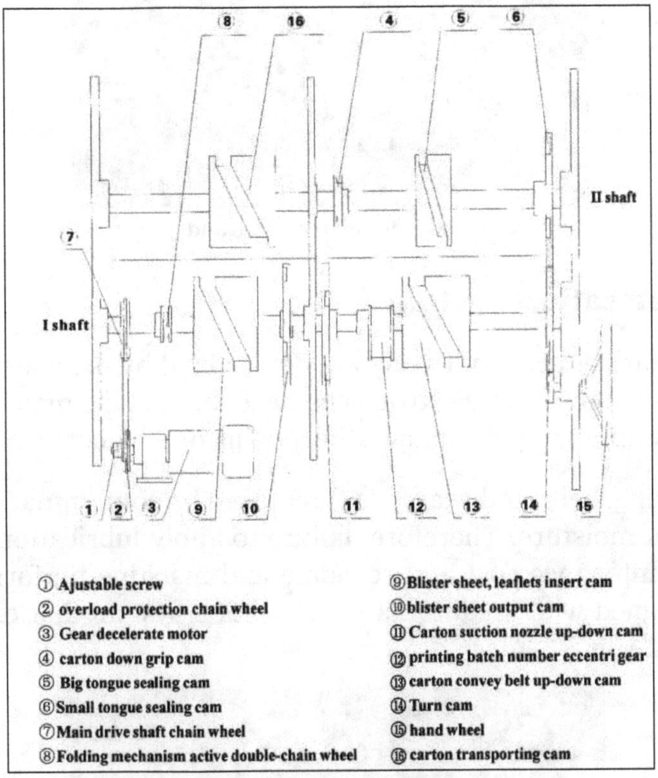

For instance, servo-driven cams are coupled on motors to produce reciprocating movement (forward and backward motion). However, it does not necessarily mean any motor can be interchangeably used as spare for another section.

Motors are manufactured with different specifications, that is, revolution per minute, phase, power supply and wattage. These features make different motors serve different functions within the carton forming machine. It is important to note the above when addressing motor breakdown. Also, the speed of these motors can be controlled by the operator via the HMI by adjusting the parameters set at the frequency converter. These motors are designed with in-built protection system such as:

- Clutch breakers.
- Torsion-overload protector.

They can prompt disengagement of the main drive motor from the gearing system thus protecting machine parts by preventing damage due to overload. The motors make it possible for the conveyor to run, cartons to move along the rail, pressing mechanism work at the carton pack tooling etc.

Motion responsible for all rotations and movements.

Automatic Lubrication System

The life-span of bearing, belts and gears are determined by the consistency of lubrication. A cartoning machine is an electromechanical device. Like many other mechanical systems, to achieve optimal performance, the moving parts must be properly lubricated.

Service life of bearing, belts and gears are threatened by contamination, high temperature, humidity and moisture. Therefore, failure to apply lubrication it may adversely impact on the maintenance cost and schedule and machine performance. Cartoning machines are designed with an automatic lubrication system that ensures processing operations are well-run.

Sections of moving parts that require lubrication.

The lubrication system is automatically prompt to supply adequate amount of good lubricant to the moving parts; though suitable lubricant should be supplied to the system during maintenance interval. This lubrication solution provides an opportunity to increase productivity and uptime. It also reduces labor costs and extends maintenance and repair interval.

Cartoning Pack Tooling and Pneumatic System

This is the section where the machine presides over the actual cartoning. Vacuum suction run by the pneumatic system comes into play when drawing blank cartons to begin the cartoning process. Carton storeroom is the section used to put cartons. It consists of back and front sections mounted with front and mid plates. The mid plate is mobile making it adjustable to the carton length. You are able to see the carton feeding entrance of a carton forming machine.

On the other hand, front bracket and back bracket are installed with a long-hole, hence can shift from right to left. This allows adjustment according to carton specifications. In this image, it shows the leaflet entrance of the cartoner:

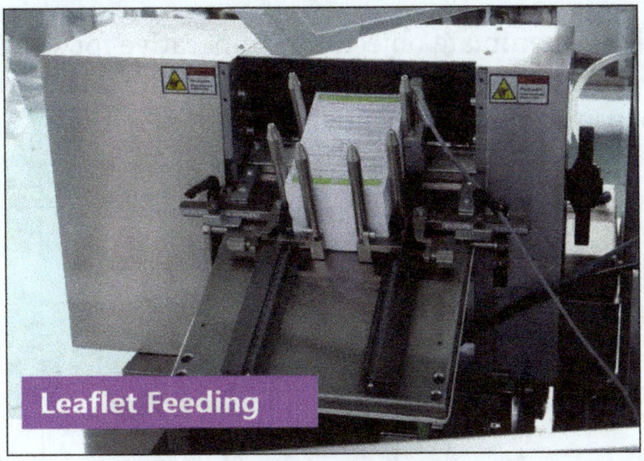

Directly in front of the storeroom is a press device. In case of blocked, damaged or deformed carton along the sucking procedure, a signal is produced by the carton press to effect an immediate stop of the machine. This is a production line interlinked with a carton forming machine.

Other parts systematically flow from feeding chambers where the product is inserted to the carton.

Leaf folder is interlinked with conveyor belt before eventually ending at the carton closing section.

An Ink-Jet Printer

A fully functional cartoning machine should be installed with an ink-jet printer, which should have writing options at least three minimum rows. Since the inventory will be inside the carton, it is wise to have relevant information about the product. For instance, date of manufacturing and expiring or any other disclaimer you ought to inform the public. This information being printed should be satisfactory because you don't what altercation arising from customers or retailers in the market.

A Safety System

Since any machine is an inanimate object, it is imperative for you to heed precautions when handling them. That is why you should request to be taken through adequate and thorough safety procedures that come with any machine.

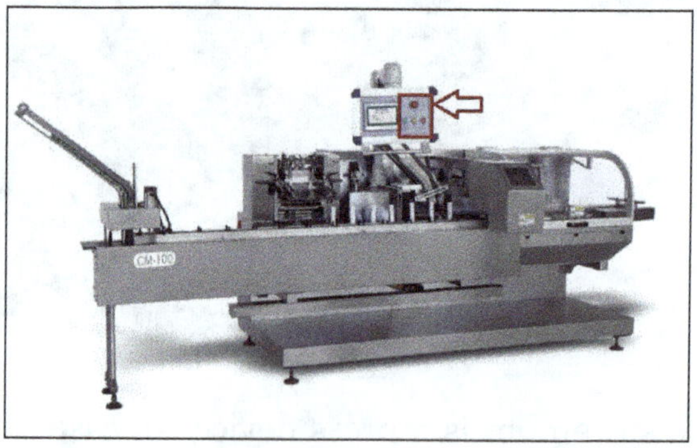

Section of essential safety components in cartoning machines.

One crucial safety is system is the emergency buttons which should be placed interval along the Cartoner sections for ease of access in case of an emergency call. Usually, the emergency button is wired from the mains supplying the machine on a Normally Closed path (NC) when operational or not. If any emergency button is in or pushed the machine should not initiate any process unless the issue is resolved and the button pulled. In that instance it becomes Normally Open (NO) meaning the power circuit is open.

However, upon noticing unusual packaging, undesired motion emanating from the moving parts; it is recommended to push any easily accessible emergency button. This will cut off power supply to the machine. In the process you would stopped a mess or losses.

Another safety measure that should be observed is proper earthing and grounding of electrical circuit running in the cartoning machine. This is to avoid unpleasant shocking from the body of the machine. The cartoning machine should be well-designed to reduce exposing the operator from unnecessary danger and risk. It should have well designated section where human operator can work comfortably without problems.

Conveyors, Rollers, Guide Rail and Carton Chute

To promote movement of cartons from one point to the other, conveyors are interlinked with rollers. They draw their motions from running motors that are coupled with gear box to determine the speed of the motors. This part is vital for operation of the Cartoner because once the product is packaged the carton moves to the next step. At the same time, it is being guided by the rail forming the runway.

Depending on the manufacturer, cartoning machine can be designed with a chute concept. At this point, fully packaged inventory is dropped from the Cartoner for further packaging into bigger boxes for storage or transport.

As a rule of thumb, always request for product manual from your manufacturer of the machine. You'll get to know every component of the machine. The number and design of parts will vary from one machine to another. This will depend on the cartoning machine model and design.

The Working Principle of Cartoning Machines

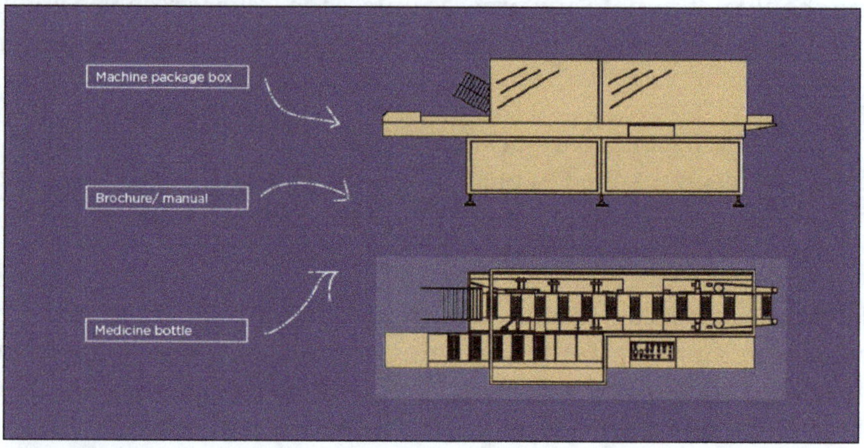

Most carton forming machines are designed and installed with three feeds or entrances:

- Machine package box entrance.
- Brochure/manual entrance.
- Medicine bottle/vial entrance.

Operational principles commence right from the feeding box to final process of package molding. Approximately four stages form the forte of the whole principles of cartoning machine.

First, a carton is set in position by the guide rail before a push mechanism by the push plate makes the carton to open up. Then the bottom of the carton is stuck to a sucker (vacuum suction) driven by the pneumatic system, from its front and rear directions to ensure that the box get to the loading bay at right angle.

At the loading area, product is inserted into the carton through mechanical effort, however, gently and precisely. Once it is filled, chain conveyor directs filled cartons along the folding spot. At this point, the cartoning machine targets ears of the cartons. The process is done simultaneously on both left and right side of the rail by rotary movement. Then the cam coupled to motors with push plates on either side of the rail does the actual folding mechanism. The next crucial step sets in, here; front covers close the tongue of the carton. Afterwards a push is orchestrated using the pushing plate, the lid is bent and resultant fastening lock as the tab gets inserted into the box.

This is the cover action. At this juncture, the degree of accuracy of the machine is assessed. It determines the good and bad package, by examining the structure of the carton before deciding on the damaged ones and those to lead carton chute.

How an Automatic Cartoning Machine Works

Gearing System of Automatic Cartoning Machine

Once the motors are powered on it initiates a mechanical movement to the main drive. It subsequently transfers it to other gearing parts through:

- Gearing axis.
- Threaded shafts.
- Support axis which are sustained by a centered plate.

It will depend on the design of the cartoning machine. This results in a reciprocating movement at the middle section of the machine. The centered plate however, is adjustable depending on the carton specifications. And at the back of the machine there is a gearing chain that ensures the centered plate follows a reciprocating motion. Carton chain transmits and intermission movements along the chain process.

Blank Cartons Section

Blank cartons are stored at the carton storeroom. At this point both carton storeroom and carton suction process form the carton unload-device. The front section of the carton storeroom is a pressing device for the stack cartons.

Apart from picking carton from the stack, the sucker is also used to absorb blocked or damaged carton between the blank carton suction point and storeroom. In most cases, a photoelectric mechanism encodes input that prompt a signal to immediately stop the machine.

Blank cartons moving to the cartoning machine system.

Feeding Mechanism of a Cartoning Machine

This will differ depending on the type of carton forming machine you are using. For instance, in bottle cartoning, there can be a turntable for aligning the products into a conveyor. It aims to link-up with the carton tooling machine, or a feeding screw.

Bottles from the turntable to the conveyor system.

The three entrances of the cartoning machine transfer all the materials to the feeding structure, which consists of feeding tank and material feeder.

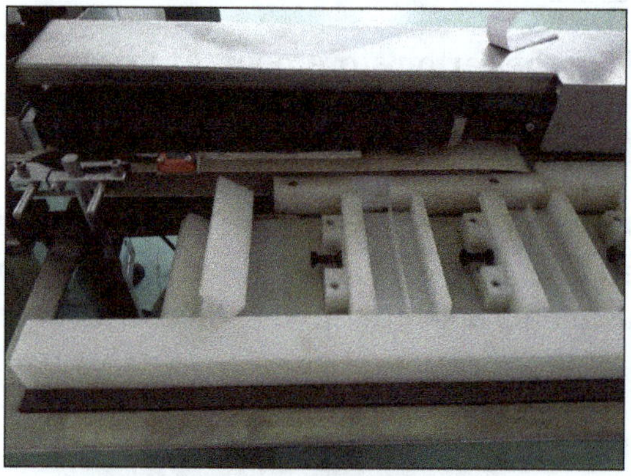

Section of the feeding system.

Folding Mechanism of the Cartoning Machine

Next is the folding mechanism. Here, leaflet folder folds up the entire leaflet while running on the leaflet conveyor. This process is followed by stacker, paper separating, suction adjustment respectively.

In the stacker adjustment, a piece of paper is placed on the feed-table and the machine-wheel set at sa position which makes the suction nozzle rise to its peak position. Then a clearance of 1mm is made between the suction hole and paper cover front edge. In fact, at the apex of the suction nozzle, both its plane and that of the stacker must be flash.

The set highest position of the suction nozzle it to enable easy reach to carton onto the bottom of the blank, thereby leaving a gap between the remaining stack the absorbed paper. On the other hand, the paper hook should be positioned at the lowest point to ensure paper suction is pulled down.

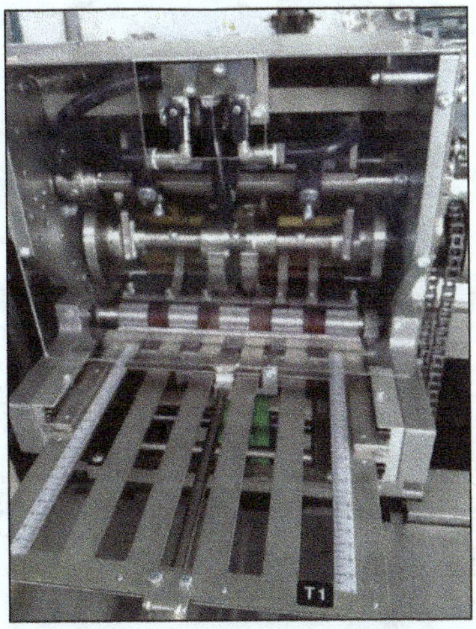

Product Insertion Stage

This is where the machine inserts the product in the respective boxes. The product is supplied to the feed device where it is taken at the bottom before correctly being pushed inside a carton. It is the push port that directs the products through a mechanical drive latitudinal or longitudinally. In an event the process stalls, an overload feedback system will alert the PLC unit on the control panel to stop the machine.

Bottles read for feeding in the boxes.

Carton-Seal Device of a Cartoning Machine

It is made up of small and big tongue carton close device. The former is on either sides of the conveyor and composed of box guide and folding pad. The carton is made to move at specific range when runs on the box track. At the same time, the close device powers the paddles and it repeats movement at a given angle range. The latter is driven by duo rotating cranks that move the plate flat. This movement results to bending, tongue and sealing.

Sealing cartons after feeding the product.

Side-seaming begins and application of hot melt-spray on both ends of the cartons while being gradually pressed along a small plane while the hot spray solidifies. Another automatically run-process of ink-jet printer prints date stamp, environmental protection labels, legible etc.

Chute of Automatic Cartoning Machine

At the carton chute a pressurized arm helps in ejecting packed carton. In addition, if the integrity of the carton is compromised, the action cover determines and decides if it is good/bad or accepted/rejected.

Key Stages in Actual Cartoning Process

We did highlight the working principle of cartoning machine and how an automatic carton forming machine works. We will majorly focus on key stages involved in cartoning process. You can notice that a cartoning process is a series of stages which lead to cartoning formation in the end. But the whole process has approximately four main stages which are considered essential in the whole process. The following diagrams capture the four main stages of actual cartoning process in the machine.

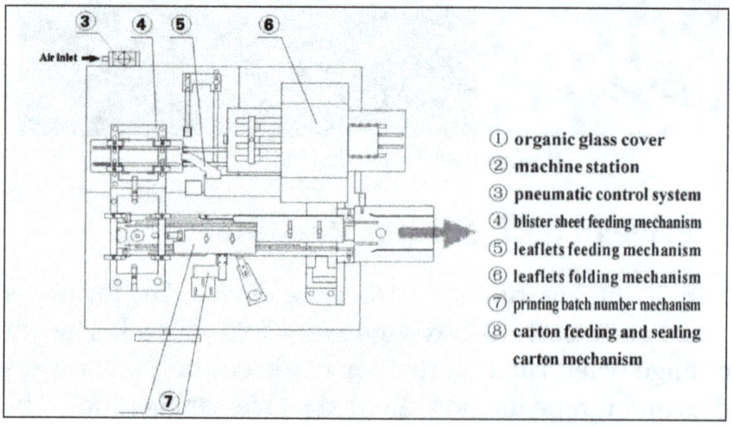

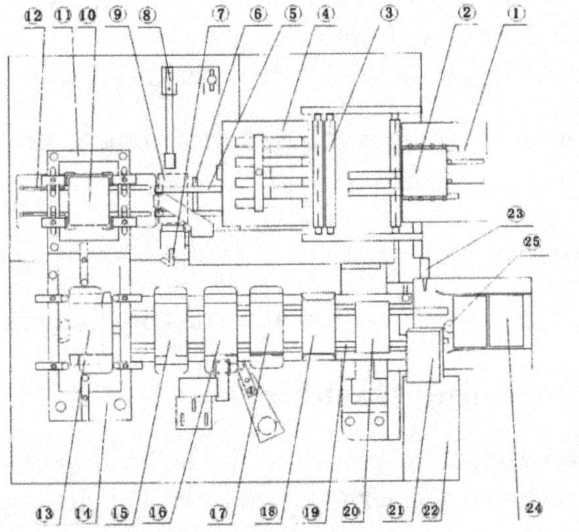

①leaflet sheet stack plate ②unfolded leaflet stack ③leaflet-floding roller ④folding leaflet adjustable plate ⑤convey belt ⑥folded leaflet ⑦leaflts waiting for inserting to carton ⑧blister sheet lognitudinal thruster ⑨blister sheet (bottle) waiting for inserting to carton ⑩blister sheet of stack plate ⑪blister sheet stack bracket ⑫blister sheet transverse thruster ⑬carton ⑭carton stack bracket ⑮carton small tongue sealing, inserting blister sheet (bottle) and leaflets station ⑯carton on printing batch number station ⑰carton big tongue pre-fold guide station ⑱carton on big tongue curve station ⑲carton convey plate ⑳carton big tongue folding and supporting station ㉑lacked leaflets reject to waste warehouse ㉒waste warehouse ㉓reject blow nozzle ㉔Finished product output ㉕finished product counter

What follows are processes of each crucial step in the carton packaging process:

- Under the Box-Action: Here, a sucker on the carton feed-mouth sticks a carton on a downward position along the cartoning line while push plate makes the guide rail clamping-box opens up the box. Concurrently, two screens move forward, from base upwards, in the opposite direction carton is stuck on the side. The mechanism ensures that the box is arrives at the loading area at right angle.
- Loading Zone: This stage is where the filling machine operates. Inserting the inventory is rather straight forward by using pressurized arm or feeding screw to drive in the product into the carton. This stage lasts at most 2 seconds.
- Folding Mechanism: What follows is the folding mechanism. Immediately after the loading stage, there is a push plate that flips up the lower end-flap to close and the next point the upper ears are shut.

  A rotational motion completely folds and presses the ears of the box while lining along the rail. This chain operation and mechanical rotary motion are thoroughly executed leaving no ear unturned.
- Cover Action: This final stage determines the integrity of the whole automatic cartoning process. The carton gets folded in tongue, and then is pushed to bend the box by pushing plate. However, the tongue is left inserted into the box. Lock

fastening proves that the cover action was well-run. As you can see, the process involves series of parts and components coordinating together. This ensures accuracy and consistency in the entire process.

Classification of cartoning equipment is based according to the Degree of automation and Product into carton.

Degree of Automation

Classification according to degree of automation has precisely two categories, namely:

Semi-automatic Cartoning Machine

Semi-automatic cartoning machine can either be horizontal or vertical cartoning machine, but an operator manually inserts the product. This invention best suits packaging in:

- Cosmetic.
- Food.
- Nutraceutical.
- Pharmaceutical industries, just to mention a few.

It has no automatic in-feeding system which makes its operations simple; however, it substantially lowers production speed as well as throughput. Furthermore, it requires many operators to get an effective production output.

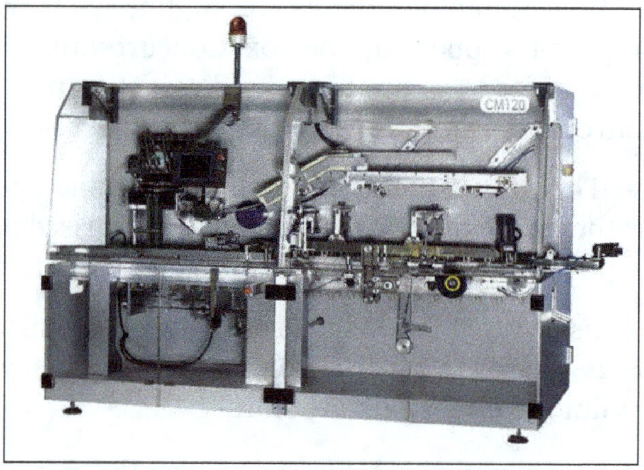

Horizontal & Intermittent Cartoning Machine.

Nonetheless, it other useful design features which enables it to handle packing blister packs, bottles, pouches, sachets and tubes in carton. More sophisticated designs can accommodate more sides accessories such as:

- Box conveyors.

- Carton magazine.
- Automatic folding machines even with devices to open and close end flaps.

For instance, semi-automatic vertical cartoning machines are designed to automatically open and close though at slow speed, but can neither erect nor fill product.

Benefits of Semi-automatic Cartoning Equipment

- Generally, its maintenance cost is low.
- They are still configurable with HMI and other safety devices for protecting the machine from damage.
- It sounds rather unequivocally simple cartoning equipment, but it can take many more upgrade of optional features such as embossing and hot stamping is possible, ink-jet printing and hot-melting, tape-sealing etc.
- Most semi-automatic cartoning machines are ergonomically designed, and are set up with guards around the equipment to ensure safety of the operator.
- It does not require highly-skilled labor as semi-skilled operator can comfortably handle its operations.
- Perfect for a wide array of batch-sizes and relatively small-medium size production businesses. Its components are adjustable making format changing easier to carry out.

Fully-automatic Cartoning Equipment

Fully-automatic cartoning equipment is a more elaborate innovation as compared to a semi-automatic one. This state-of-art machine has been built in manner that reduces too much human involvement when it's operational. Apart from supplying packing materials at its entrances, all other tasks of:

- Carton erection,
- Ear folding,
- Product insertion,
- Carton closing,
- Hot spraying,
- Date stamping,
- Ink-jet printing.

Fully automatic cartoning equipment for bottles.

Even the product chute is exquisitely executed by the fully-automatic cartoning equipment. Ideally, fully-automatic cartoning machine captures and defines best the working principles of an automatic cartoning machine. All packaging processes are solely controlled, managed and regulated by control unit of the machine.

Its accessories units such as the ink-jet printer, folding systems, hot melt gluing are harmoniously coordinated to conveniently run the packaging exercise at the same time accomplishing all possible requirements like brochure insertion and many more. Moreover, fully-automatic cartoning machines can check on the integrity of the carton after the process and decide whether it is damaged or good.

Automated feeding system.

Here, we have different types of modes found in fully-automatic cartoning machine, namely: Continuous and intermittent modes.

Continuous Automatic Cartoning Machines

They are also referred to as high-speed cartoning machines. That is why they are specifically designed for high speed production lines. This is mainly for regular shaped and solid inventories that require no special handling provisions. Its chain process runs on guide rails; conveyors allowing for progressive execution of each packaging procedures including product insertion.

Intermittent Automatic Cartoning Machines

On the other hand, intermittent automatic cartoning machines stop to allow boxing cycle. Compared to continuous automatic cartoning machine, intermittent cartoning machine is slower by 150 times.

Both continuous and intermittent cartoning equipment are designed to pack products laterally into cartons and seal by lock fastening during flap insertion or hot-melt gluing.

In both, product insertion is orchestrated by either electronic or synchronized mechanical device. Having exhausted much about classification according to degree of automation, it is best we proceed to the second classification based on product into carton.

Product into Carton

This class forms the type of commonly used cartoning machines in most industries. It is based on how product introduction is done into the carton. There are two types of cartoning machines under this class. The following is the list:

- Vertical Cartoning Machine.
- Horizontal Cartoning Machine.

First, both the vertical and horizontal cartoning machines can either be fully-automatic or semi-automatic by design though they differ in their mode of production insertion.

Vertical Cartoning Machine

Basically, a vertical Cartoner products are introduced into the carton vertically directly from in-feeding system and sealing machine into folding box. Hence, the name vertical cartoning machine attributed to its mode of product insertion.

Vertical Cartoner is also designed to pack a large of range of product specifically ideal for handling bottles, jars and vials. Therefore, it is mostly preferred in cosmetic, food and pharmaceutical industries. Most vertical cartoning machines come with vast flexibility and compact dimensions as main features. This makes them adaptable thus they can meet most industrial challenges.

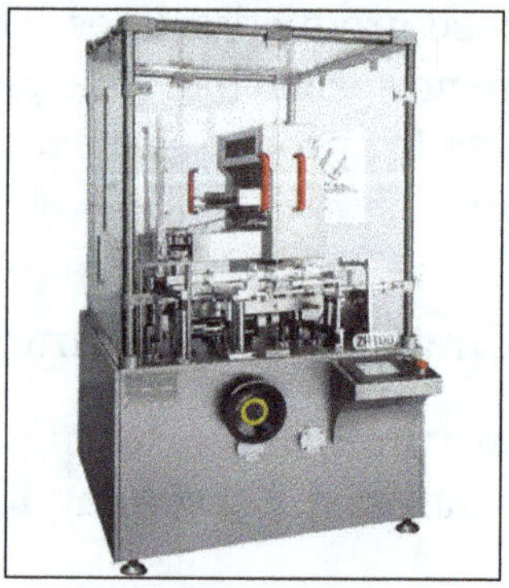

Compact Cartoning Machine for Blister.

A combination of simple format part changeover and modern technology has made it possible to achieve a vertical carton position for product feeding. In fact, statistics has proven that vertical product insertion method records low risk of product damage. At the same time get rid of lengthy delays of in-feeding systems.

Benefits of Vertical Cartoning Machine

When you opt for vertical cartoning machine, you should expect to enjoy the following key benefits:

- Features simple and robust format changes. For instance, on a curve, its rail requires no adjustments.
- Can adopt two working principle as intermittent and vertical cartoning equipment.
- Provides accessible work stations for operator.
- Open structure makes it easy to clean.
- They can be designed and manufactured with HMI thus simplifying its operations to one operator.
- Can be equipped with additional modification of transport system.

Horizontal Cartoning Machine

This is a similar innovation to vertical cartoning machine, but the stark difference is that a horizontal Cartoner shapes cartons from stack of blanks. Basically, it introduces products by lateral drawer, inserting or pushing and ultimately closes the carton.

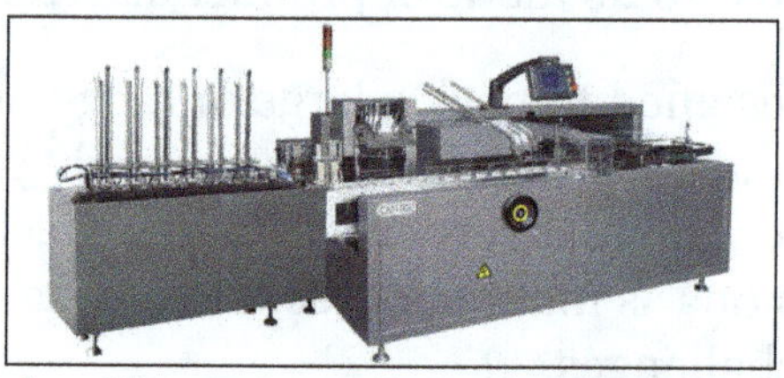

Horizontal and Intermittent Cartoning Machine for Sachet.

Usually, horizontal cartoning machine is interchangeably referred to as automatic cartoning machine. It can either be manufactured as continuous or intermittent motion machinery. This makes it suitable of achieving high speeds and suitable for high demand production applications. The following are some of standard features you should consider in a horizontal cartoning machine:

- It should either have auto or manual loading.
- Its conveyor network or flow should be adjustable by design.
- Expect simple accessories such as hot melt gluing or tuck-in flap.
- Tuck closure.

Horizontal cartoning machine is ideal to pack products with stable base even with one or two items in additional. It is commonly used in packaging for food industries: cake, coffee, packs, frozen pizza, cereals etc.

Canning Equipment

Canning is a process whereby selected raw materials are prepared as for the table, packed in tin or glass, or plastic containers capable of being sealed airtight, heated sufficiently to destroy spoilage and pathogenic agents within the container and cooled rapidly to prevent over cooking of the product.

Can Making Equipments

- Can body Reformer: The can bodies, in most of the cases are supplied to the distant canners in collapsed form. So as to reduce space and easy transportation they can bodies are shaped into cylindrical form using in a can body reformer.

Can body Reformer.

- Flanger: The cylindrical shaped can bodies are flanged in the can ends by the flanger. The flanging has two flanging dye which have flanging of grooves.

Flanger

- Seaming Machine: The Seaming machine helps in closing one end of the can. It has a base plate for keeping the can that can be raised when the can body is raised the upper end of the can on which lid is kept comes in contact with the top plate and simultaneously two seaming rollers of first and second operate. Thus closing the can.

Seaming Machine.

- Can Leakage Tester: The seamed cans are tested for leakage using this machine. The cans are immersed in water and pressures are noticed, then it is understood that seaming is not proper.

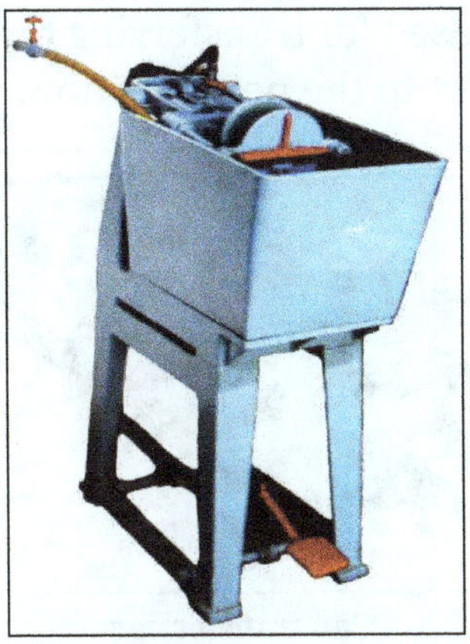

Can leakage tester.

- Flange Rectifier: Sometimes the cans subjected to flanging will be bent, deformed or crushed. These types of cans are rectified by flange rectifier machine.

Flange rectifier.

- Can coding Machine (Embossing Machine): The machine is used for embossing the code, which contain the name of the product, name of manufacture (code) and date of production.

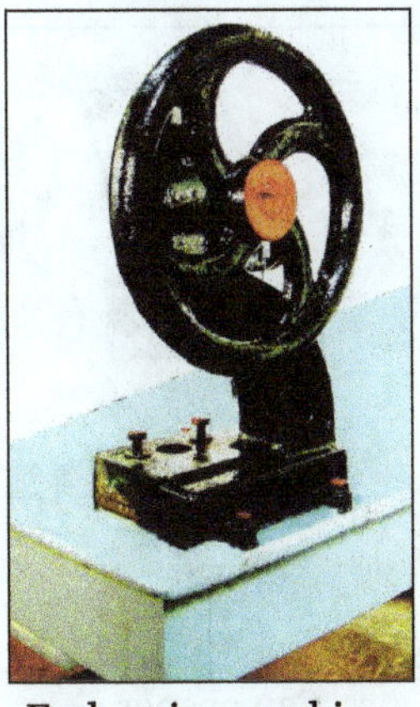

Embossing machine.

- Transfer Car: The car is used for transferring the cooker tray (basket/plate type) taken out of cooker/retort to the next position.

Can transfer car.

- Packing table with conveyor: This equipment consists of a packing table and a conveyor belt. It rationalizes canning operation and greater production efficiency. The packing table is made of stainless steel to resist corrosion.

Packing table with conveyor.

Double Seam Examination

- Can Opener: The special manual can opener is a heavy-duty unit commercially available from most container suppliers. It is adjustable to the can diameter and removes the center panel at the cover without damaging the seams or the body. The use of conventional house hold can opener should be discouraged because it leaves a double seam which is not only difficult to tear down, but may be distorted.

Can opener.

- Can Vacuum Tester: This instrument is used for the measuring vacuum with in a closed can. The cushion of the tester is pressed hard on the lid, thus making its needle tip pierce the can. The measurement at vacuum appears on the dial of the instrument.
- Seam Micrometer: The Seam micrometer has a standard micrometer barrel incorporated on to a suitable stem, which has been designed specifically for measurement of double seam dimensions. The micrometer is placed over the surface to be measured by balancing the micrometer with the index finger at 90-degree angle to the surface. The micrometer is lowered on to the surface touches the surface. The adjustment barrel of the micrometer should then be used to bring the stem in contact with other side of the surface without exerting excess pressure.
- Countersink Gauge: The dial indicator type counter sink gauge is probably the best. However, there are less expensive instruments available. All are used to measure the countersink depth from the top of the seam to the lowest point adjacent to the countersink wall.
- Nippers: Standard no.5 nippers are available in most hardware stores. Nippers are used to tear off the strip of cover left by the can opener.
- Seam Saw: A cross-section of the double seam can be cut with a fine metal saw.
- Seam Projector: A seam projector is very effective for direct measurement of the seam overlap. A small cross section is first removed from the double seam area of the can. It is then clamped into position and calipers in the base of the projector permit direct measurement on the projected image of the overlap, cover hook, and body hook. The projector allows inspection of the total configuration of the seam at the same time. It should be understood that this unit is limited to observing a cross-section of the seam. It does not eliminate the need for seam teardown to make a full and complete evaluation.
- Seam Scope: The seam scope is a pencil like magnifier with an attached light source. It contains a visible scale by which the length of the body hook, cover hook and overlap can be read directly in thousandths of an inch on the exposed cross-section of the double seam. Its uses and limitation are similar to that of the seam projector.
- Head Space Scale: It is used for measuring of empty space left out in the filled canned products.
- Refractometer: Used for testing the concentration of brine solution and brix concentration of syrup used as a filling medium of fish and fruits canned products respectively.
- Vernier Calipers: This instrument is used for finding dimension of can.

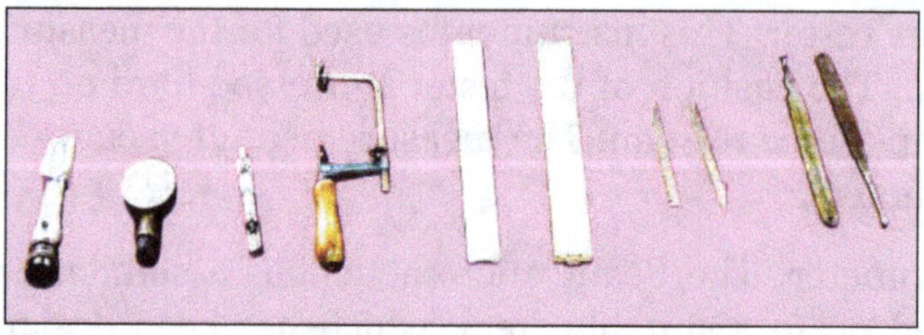

Refractometer, Vacuum tester, seam micrometer, Can cutter, Counter sink gauge and Tone tester.

Home Seamer

Hand Operated: Adjustable to Different Can Sizes

This is manually operated closing machine for seaming open-top cans. It is portable and recommended for a laboratory or for small company.

Specifications

- Range and can size: 2 1/8" - 4 1/16" dia. or 2"-5 ht.
- Capacity: Cans per m^2.
- No. of Revolution of hand wheel to close one can: 30.
- Floor space: 21 x 14.
- Overall ht.: 53"
- Gross Wt.: 300 lbs.
- Cubic Measurements: 12 cft.

Home Seamer (Hand operated).

Type: Irregular Can Double Seamer

(For oval, square, rectangular and other irregular shaped cans). This is a semi-automatic all-round machine for closing oval, round, square, rectangular and other irregular shaped cans. The operator adjusts the seaming head, places on the lifter a filled can with a cover on and depresses the foot-pedal. The lifter spindle will ascend and fix the can in the seaming chuck. A further depression of the pedal will cause the seaming head revolve and by the alternating action of first and second rolls, seaming is affected. It is a belt driven machine, but motor driven is also available.

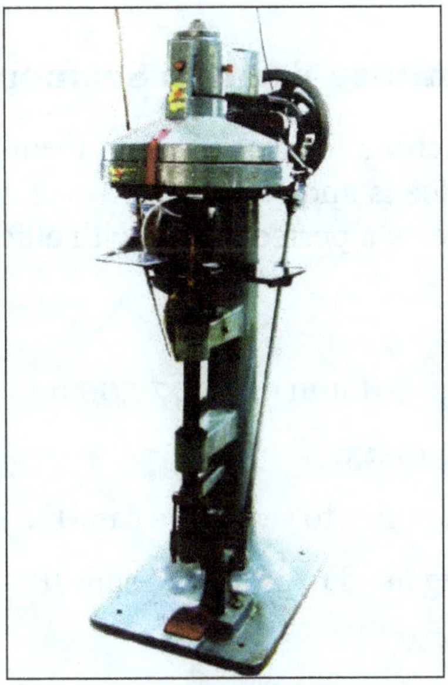

Irregular can double seamer.

Specifications

- Range of can size: 3"-8" dia. or 2" -10" ht.
- Speed: Cans per minutes: 10-15.
- No. of Revolution of driving pulley to seaming one can: 20.
- No. of Rev. of seaming head to seam one can: 10.
- Required horsepower: 1 H.P.
- Floor space: 48" x 28".
- Size of base: 33" x 23".
- Over all ht. : 72".
- Gross wt.: 1200 lbs.
- Cubic measurement: 68 cft.

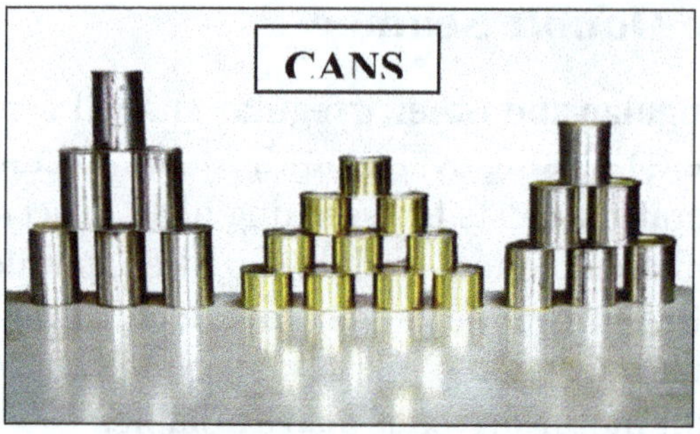

No.1 Tall 8-Oz 1 lb Jam.

Model-5-Automatic Sanitary Vacuum Seamer

This is a fully automatic machine for closing round cans in a vacuum chamber. The machine has cover and can-feeds and No can-No cover mechanism. It has a superior seaming set-up capable of giving a perfect result with efficiency.

Specifications

- Range of can size: 202-401 mm dia., 107-500 mm ht.
- Speed: Cans per min. 40-45.
- No. of Rev. of driving pulley to seam one can: 15.
- No. of Rev. of Seaming head to seam one can: 18.
- Required power: 1.5 KW.
- Floor space: 1200 x 1450 mm.
- Size of base: 650 x 850 mm.
- Gross Wt.: 1480 Kg.
- Crate Dimensions: 1300 x 1600 x 2000 mm.
- Cubic measurement: 4.2 m^3.

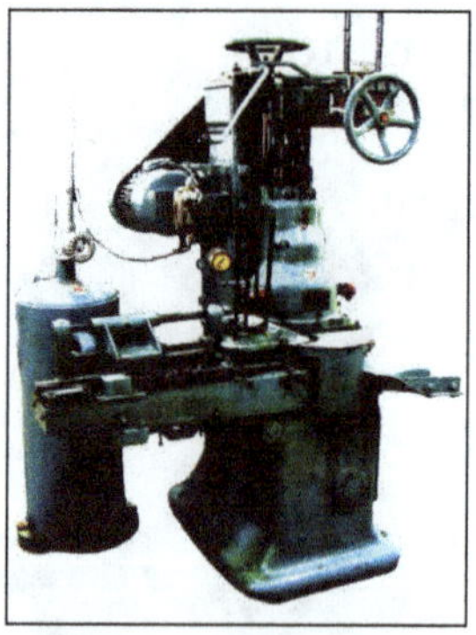

Automatic vacuum seamer.

Semi automatic vacuum seamer.

Principle of Double Seam

The double seam is made in two operations. To begin with, the first operation rolls the flange of the lid under the flange of the body of the can, and then the second operation rolls finish off the seams by tightly pressing together and rolling up all five layers of tin plate. In this machine, the seaming is made up with two or four grooved rollers, which are brought into operation by a can or eccentric arrangement fitted into the hand.

The first operation roller, which comes in contact with the cover, has a narrow groove, and the second operation, which follows the first, has a wide groove. Inside, the curl of the cover is a rubber compound which is applied in a liquid state and dried out, leaving sufficient rubber or compound for jointing. When the seam is complete, the edge of the body is pressed into this material. The first operation roller folds the cover and flange together and the second operation flatten out the metal and complete the seal. The lid is seamed while one can is under vacuum.

Setting for a Correct Seam

In setting up a double seaming machine, the first thing to do is to check the chuck and the seaming rollers to be sure they are in good condition. Next step is to bring the base plate up to the top position. The setting of this plate is the adjusting of the distance between the face of the base plate and the top edge of the chuck. This is done by placing a can on the base plate and adjusting up or down until the flange of the can is just level with the top edge of the chuck i.e. distance equals height of the can minus the thickness of the chuck.

Now, turn the machine round slowly and bring in the first operation roller and make sure that the distance between the top edge of the chuck and flange of the roller is not less than 0.002 inch. Now, turn the machine further until the second operation roller is in contact and allow not less than 0.015 inch of clearance.

To facilitate the setting of the rollers, the canner is usually supplied with two wires, a thick one for the first operation and a thin one for the second operation. The machine is turned round until the first operation roller is in full operation and the roller is then adjusted to a point where the wire will pass firmly between the contour of the roller and the face of the chuck. The same process is repeated with the second operation rollers and the thin wire.

When the adjustments are all made, make sure that all locking nuts or bolts are tight and turn the machine one complete cycle by hand before seaming a test can under power. The seam should first be measured on the outside to check thickness and length. The hooks and double seam are measured with a micrometer. There are some standard limits of seam dimension.

Maintenance

The machine should be kept well-oiled and greased. The seaming rollers or seaming roller bearing should be greased not less than once per shift or hours. At the end of the shift, the machine should be thoroughly washed with hot water and then with steam and the setting of the rollers should be tested while the machine is still warm. After this is done, oil should be rubbed over the metal faces to prevent rusting. Not advised to exceed the speed specified by the manufacturer.

Pressure Process Retort

Horizontal Retort

Two models such as No. 48 and 510 are available. This unit is used in a cannery with a steam. This equipment is fitted with Pressure Gauge, Thermometer and Safety Valves.

Model	48 horizontal	510 horizontal
Inner diameter	940mm / 291 /8"	910mm / 357 / 8"
Length of body plate	1220mm / 48"	1520mm / 60"
Thickness of body	6mm / ¼"	8mm / 5 / 16"
Thickness of lid and end	8mm/ 5 / 16"	9mm / 3 / 8"

Horizontal retort.

Astoria Type

A retort is a steel tank in which crates containing filled cans are placed for processing and cooling. It is fitted with a cover without any pressure leak. For large-scale production and more efficiency, horizontal retorts are preferred than the vertical retort.

Retort Equipment

Indicating Mercury-in-Glass Thermometer

Use of two such thermometers on each retort is advised – one so placed as to serve as official guide to the operator in his control of temperature of processes and the other intended to serve as a check instrument located adjacent to temperature recorder bulb.

Each thermometer should have a temperature range of not more than 100 °F, For example: 170 °F to 270 °F. The thermometer should have a pressure scale corresponding to the temperature scale at the altitude above sea level at which the retort is located. The relation between the temperature and pressure scales at various altitudes is provided in a table. The pressure scale of all thermometers should be adjusted in correspondence with these data.

Pressure Gauge

It should be graduated in one pound division and have a range up to 30 lbs. The gauge should be connected to the retort by means of a gauge siphon.

Over pressure retort.

Valve Controlled Vents

In general, the venting efficiency is better when the resistance to flow through the venting system is decreased. This resistance is minimized with the larger size fittings and gate-type valves rather than the globe valves. Vents should be controlled by quick opening gate or plug cock type valves, which permit unrestricted discharge of air from the retort during the come-up period.

An air-steam mixture in the retort will cause a non-uniform heat distribution, and may cause a lowering of rate of heat transference to the containers. The specified time of process should not be considered as beginning, (even though the proper temperature has been reached) until temperature and pressure instrument on the retort corresponding readings.

Processing in Open Kettle

Fruits, including tomatoes and other food with sufficient acidity, may be safely processed in open kettle at the temperature of boiling water.

Jacketed Steam Kettle

Both rotating and stationary types are available. The inner kettle will be made of cast-Iron, stainless steel or gun metal according to the purpose for which the container is to be used. It's containing capacity ranges from 360 to 370 liters. Its other accessories are pressure gauge, safety valve, stop valve and drain cock.

Jacketed steam kettle.

Semi Automatic Corer and Sizer

This machine is used to core, peel and size pineapples or apples. The cutter, core-tube knives are made of non-corrosive acid resisting material. It is very easy to install and operate.

Specifications

- Range of can size: 80-105 mm dia.
- Speed-Fruits per min.: 15-20.
- No. of Rev. of pulley: 550 – 600.
- Required power: 0.2 KW.
- Floor space: 1800 x 440 mm.
- Overall height: 1370 mm.

- Gross Wt.: 220kg (Approx).
- Cubic measurement: 1.65 m^3.

Pineapple corer.

Pineapple slicer.

Model "O" Sanitary Vacuum Seamer

This is a semi automatic machine for seaming round cans in a vacuum chamber. As it is quite simple in operation and adjustment, it is recommended for use in a small cannery or a laboratory. Model 5-P Vacuum pump, vacuum tank and their pipes. The Seamer alone can be purchased without accessories. Normally this machine is operating by belt drive, but meter drive can be available on demand.

Specifications

- Range of can size: 21 / 8" -6 3/16" dia. and 2" -62 / 16" ht.
- Speed Cans per min.: 10-15.
- No. of Rev. of driving pulley to seam one can: 18.
- No. of Rev. of Seaming head to seam one can: 18.
- Required horsepower: For Seamer-1 H.P; For Pump - 5 H.P.
- Floor space: 38" x 28".
- Size of base: 32" x 28".
- Overall ht.: 80".
- Gross Wt.: 1625 lbs. (approx.).
- Cubic measurement: 80 cft.

Table: The standard limits of seam dimensions (inches).

Can diameter	Nominal substance	Dimension		
(inch, 1/16")	lb. (per base box)	Thickness (1/1000")	Length (1/1000")	Hooks (1/1000")
211	85	55-57	117-125	75-85
300, 301	90	56-58	117-125	75-85
307	95	58-60	117-125	75-85
313, 401	100	60-62	117-125	75-85
401, 502	108	63-65	117-125	75-85
603	112	65-67	122-130	80-90

Thickness measurement can only be considered in relation to the actual substance of the body and end examined. Allow 0.002 inch for each 10 lb. deviation in substance of body plate, and 0.003 inch for each 10 lb. in end plate. Seams made from lacquered plate will be about 0.003 inch thicker than those made from plain plate.

Stretch Wrap Machines

Stretch wrapping loads keeps them from moving and shifting during transportation which is a major source of damage. There are certainly other ways for loads to be damaged during shipment, but stretch wrapping provides a measure of protection that increases the odds of "as made" condition arrival. In turn, that means lower returns, administration duties, credit due to receiver and rework costs – not to mention happier customers and colleagues.

In the 1970's, stretch wrapping machines established themselves as the preferred packaging method for unitizing goods for shipment. Meaning that any company that deals with production, reselling, or warehousing of products probably has some familiarity with stretch wrapping. To make the best decisions about stretch wrapping machines and film, however, it is useful to understand how the mechanics of stretch wrapping work.

Stretch Wrapping Begins with Pallet Loading

Before the stretch wrapping process can begin a load must first be assembled on a pallet. Once the load is on the pallet, the whole ensemble (pallet and load) is stretch wrapped together. Assembling the load on the pallet is an important step. Load placement on the pallet is key. The load should not overhang the sides of the pallet or be more than four inches inboard. The pallet itself also plays a role and should be in good condition.

It is important when stretch wrapping not to skip steps that secure the load the pallet. One options is to wrap all the way down to the bottom of the pallet. However, this runs the risk of film punctures from forklifts or pallet jacks when picking up the load. Such large punctures to the film can propagate up the load, causing a loss of containment force and increasing the risk for load failure.

Pallet Grip is a great solution to lock the load to the pallet and avoid film punctures. The bottom few inches of the film web are automatically rolled into a tight cable at the end of the wrap cycle. The cable is driven down just below the top deck boards of the pallet. This strong cable locks the load to the pallet and leaves enough room to clear forks and avoid the puncture problem.

Three Ways to Apply Stretch Wrap

There are three ways to apply stretch wrap to a pallet of products: manual, semi-automatic stretch wrapper, or high speed automatic stretch wrapper.

- Hand wrapping, as the name implies, relies on people to manually stretch wrap a load. A worker secures the film to the pallet and then walks the film around the load. Pulling a heavy roll of film around a pallet is wrapping is physically demanding – especially if done right. Imagine the awful dizzy feeling in a hot warehouse. Hand wrapping generally has the highest material cost because there is no consistency of film usage and no substantial pre-stretch. It is also very difficult to wrap a safe-to-ship load consistently.
- Semi-automatic stretch wrappers come in two different machine styles – turntable and straddle. The load is place on or in the stretch wrappers wrap zone and stretch film is applied. Turntable models will rotate the load on a turntable while straddle style models have a wrap arm that will move around the load and apply film.
- High speed automated stretch wrappers are similar to semi-automatic styles in that a machine performs the actual stretch wrapping of a load. However,

automatic stretch wrappers typically include infeed and exit conveyors to automatically cycle loads through the machine.

There are a number of features that can be added to semi-automatic machines that offer some increased automation. For instance, automatically cutting and clamping the film and a remote start lanyard allow fork truck operators to remain safely buckled in their seats on the trucks even while wrapping loads on a semi automatic machine.

For more help choosing the right level of automation and machine style check out our Stretch Wrapping Fundamentals pages. That's actually the critical factor to consider – how many loads/day may not capture the surges which often occur late in the day as shipments are picked up. So loads/hour is an important factor.

It's all About Containment Force

No matter which type of stretch wrapper you use, containment force is the most important things to get right. Containment force can mean the difference between a load arriving at its destination intact and arriving damaged and unsaleable. So what is containment force?

Containment force is what holds the load together. Officially, it's the wrap force times the number of film layers. Wrap force is simply a measure of how tightly each layer of film is wrapped around the load. Film layers are the number of wraps applied to a load at a specific point. There is no magic formula to determine the right amount of containment force. It is done through trial and error. Wrap, test, change and repeat until you can safely say loads are arriving without damage.

Ideal Stretch Wrap Machine

Stretch film is engineered to be elongated, which means that once it is stretched the true characteristics of the material emerge. Because of its engineered properties, the film has much higher puncture and tearing resistance when stretched. It also has greater load holding ability.

Stretch-wrap machines are cost justified on material, labor savings, and load unitization improvements with the majority of machines paying for themselves in one to two years. In terms of material savings, consider that today's high performance machines pre-stretch the material by 250 percent or more. This means that every foot of stretch film when off the roll, covers two-and-a-half-times more area than if the material was not stretched at all.

Compare "percent stretch" to the money a motorist would save over a year's time by improving an automobile's mileage from 10 miles per gallon to 35 miles per gallon, a 250 percent difference. Stretching the film offers that same type of saving — going farther at less cost. The material savings is highest when an organization upgrades from manual wrapping, only 10 to 20 percent stretch, or from an older stretch wrapping machine to a new semiautomatic or automatic machine capable of 250 percent or higher pre-stretch.

At one end of the labor spectrum, manual hand wrapping involves 100 percent direct labor. Semiautomatic stretch wrapping involves less direct labor – a worker attaches the stretch wrap to the pallet load, starts the machine, and then cuts the film tail at the end of the process. During the wrap cycle, the operator is free to attend to other tasks. Fully automatic wrapping equipment requires virtually no direct labor in the process. After the load is delivered via conveyor, the system automatically wraps the load, cuts and weighs down the film tail, and transports the wrapped load to a staging area ready for pickup.

Managers should also factor into the justification equation the productivity gains from faster stretch wrapping and the operational advantages of freeing workers for more value adding duties. Machine wrapping can also offer a decrease in goods damaged during shipment and fewer worker issues, including falls, sprains, and repetitive motion injuries incurred during manual wrapping. A less tangible but still important consideration is the improvement in customer and channel partner satisfaction from the quality and consistency from machine wrapped pallet loads.

Types of Stretch Wrappers

Turntable and rotary tower are the two basic designs of stretch wrapping equipment. Each of these two categories is available in either semiautomatic or fully automatic versions. Many manufacturers offer a hybrid of these models called portable automatics that combine the economy price of semiautomatic models with the automated film-tail treatment of fully automatic machines.

Turntable Key Features

Pallets are processed on the machine's turntable. As the turntable revolves, the pallet load is wrapped according to a predetermined cross-spiral pattern. Wrap pattern maximizes holding power while minimizing material use. Turntables come in two basic configurations, low profile or high profile. Low-profile machines are ideal in situations where pallets are primarily loaded by pallet jack – the load is rolled up a ramp to the turntable, which is only about three inches off the floor.

High-profile turntables are designed for loads delivered by forklift. The platform on high-profile machines is 10 to 14 inches above the floor. The added height – above the level of forklift wheels – helps protect the equipment from damage due to forklifts rolling over the turntable. High-profile turntables are square shaped with rounded corners, which allows the forklift to approach closer and more easily center a load.

When specifying a turntable machine, make sure it can handle the weight of the heaviest load to be wrapped. A key component of a turntable machine is its load-bearing base frame, which should be fully constructed of structural steel components running the full length of the frame. This feature gives the machine more rigidity and better stability on uneven surfaces.

Rotary Tower Key Features

When using rotary tower machines, the load is wrapped on the floor or on a conveyor. A rotary arm travels around the load, wrapping it in a cross spiral pattern. Rotary tower machines are ideal for unstable loads that might come apart if rotated on a turntable or for extremely heavy loads because the pallet can be placed on the floor under semiautomatic wrappers. Customers also use semiautomatic rotary tower machines in areas of the plant requiring frequent wash downs because the floor near the machine is accessible.

Rotary tower machines can be placed over a weigh scale so that the load can be weighed and wrapped at the same time. They can also be placed over conveyors and other equipment for greater operational flexibility. During operation, rotary tower machines have a larger footprint than turntable machines, although when the rotary arm is not in operation they take up less space. When choosing semiautomatic or fully automatic equipment, look for structural steel construction, adequate safety features, state-of-the-art control, and easy access to service components.

Shrink Wrapping Machines

Among the specialised plastic packaging systems, shrink packaging or commonly known as shrink wrapping and stretch wrapping are very common and widely used. Though there are some apparent similarities in the two systems, in overall analysis, they are considered to be totally different in terms of material and operation.

In the case of shrink wrapping, shrink film is used as the basic material and heat forms an important part of the operation, whereas, in the case of stretch wrapping, stretch film is used as the basic material and no heat is applied during the operation. Shrink Wrapping is done in 3 or 4 stages, namely:

- Wrapping (sleeve wrapping or over-wrapping).
- Sealing (necessary only for over-wrapping).
- Shrinking (with application of hot air).
- Cooling.

Stretch wrapping is done only in two stages, namely, wrapping and sealing (most of the time even without a sealer). However, in many packaging applications, both the systems are considered as alternative to each other. These systems are mainly used for unitisation, but sometime they are also being used as primary packaging system. Both the systems are used for bulk packaging as well as retail packaging.

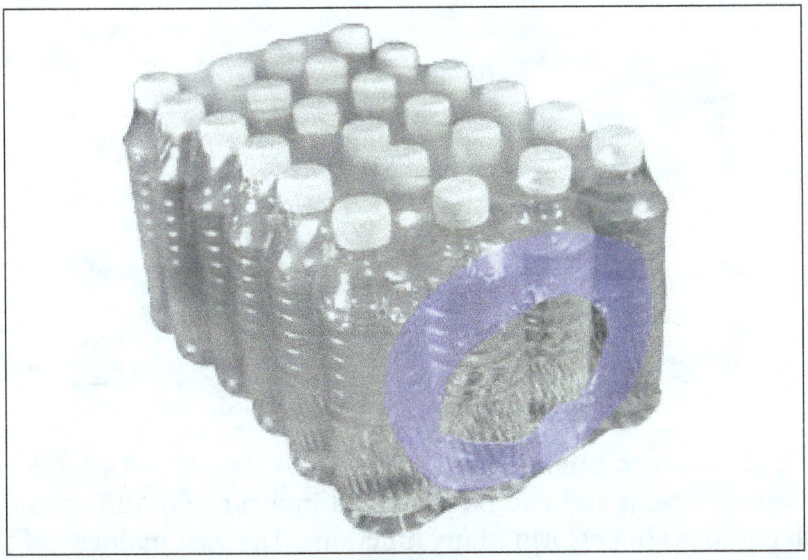

Shrink Pack

In fact, both the systems are now being used for some major applications like sleeving for labelling on various containers or sealing, besides wrapping. For both the packaging systems, material plays an important role, because without the correct material, proper shrink or stretch wrapping may not be possible. It is therefore, absolutely essential for anyone involved with the subject to either learn about the technical specification of the materials or work very closely with the material supplier, preferably both.

Plastics used for Shrink/Stretch Wrapping

Shrink and stretch wrapping can be quite complex in their structure. Most of the packaging films that are used for shrink and stretch wrapping are from the polyolefin range. These are materials produced from oil based chemicals by what is called a polymerization process, which basically means getting the right molecules and atoms to club together in a way that is required or desirable for a particular application. The most common plastic materials are polyethylene, polypropylene and poly vinyl chloride.

Polypropylene is comparatively less used in shrink and stretch wrapping, because it is slightly harder than the other commonly used materials. It has a higher melting temperature and is less stable when shrinking. However, many over-wrapping machines use polypropylene and some can be put through a shrink tunnel to give a slight tightening effect.

PVC is a dense material. As most polymers are sold by weight and there has been ecological pressure in Europe and America against its use, sometime use of PVC is restricted.

However, it is still considered to be a common material. When clarity is an important selection criterion, particularly for consumer packaging.

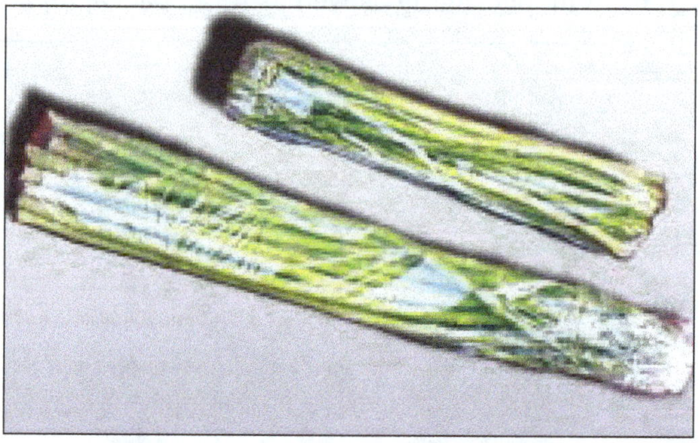

Fresh Asparagus - Stretch Wrapped.

Polyethylene is the most commonly used material for shrink and stretch wrapping because it is relatively cheap and can be produced in a range of different densities and modified with additives to perform many functions. The vast majority of shrink film is LDPE and some of the more sophisticated films have blends of LLDPE as well. Sometimes a little quantity of HDPE material is also added.

For selection of plastic material, besides type of plastics, the yield of the film is also important to be considered from the economy point of view. The yield means the area obtained or number of square meters of film converted from a kilogram of material of a given thickness or gauge.

The gauge may be expressed in micron (0.001 mm), mil (an American terms for thousandths of an inch i.e. 0.001 inch) or simply gauge, which is the old British system where 100 gauge equals one thousandth of an inch. The co-relation of both the systems of measurement of thickness is 25 micron (or 0.025 mm) equals 100 gauge (or 0.001 inch).

For shrink film, the next important factors considered are the shrinkage and the slip of the film. Shrinkage means the percentage shrink in the machine direction, i.e. along the reel of the film, and also in the transverse direction, i.e. across the reel of the film. The slip can be of different types – high, medium or low depending upon how much slippery property in the film is required from the operational point of view. Usually, low slip is desirable. For shrink wrapping small packs at high speed, particularly for consumer products or display purposes, PVC or specially modified Polyolefin may be used. The "high shrink films" are crystal clear but generally expensive.

For stretch wrapping, majority of films are modified linear low density polyethylene (LLDPE), often three-layer extrusion with "tackifier" added to make them sticky on either one or both sides. Some PVC films are also used.

Manufacturing of Shrink Film

Almost all plastic films shrink to some extent. In most packaging applications, dimensional stability is desirable and shrinkage is considered to be detrimental. However, in some packaging application, controlled shrinkage is deliberately used for certain objectives to be fulfilled. For example, shrink film is used to pack products with a tight pilfer-proof wrap over carton and boxes, so as to utilize number of packs or items on pallet, etc. A variety of shrinkable films have been developed to meet the requirements of many different end-use applications.

Heat-shrinkable film is made by stretching or orienting a conventional film at a temperature close to its softening point (i.e. Tg) and then quenching or freezing the film in the oriented state. The film that undergoes a special stretching and cooling process, causes an orientation in the film and introduces frozen-in shrinkage stresses, which can be given hot air treatment or infra-red radiation. Figure shows a schematic representation of a single long-chain polymer molecule in the relaxed and considerably curled state that is characterised in an un-oriented film.

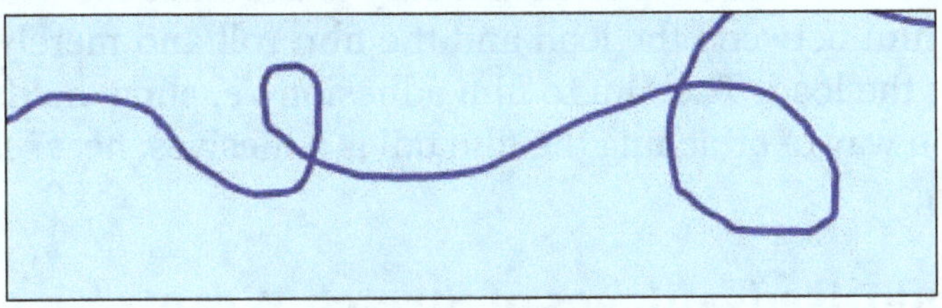

Polymer Molecule.

If the same film is stretched i.e. uni-axially oriented in one direction while it is heated, randomly twisted and intertwined molecules line up as shown in figure. However, if the film is stretched not only in the machine direction, but also in the transverse direction, there will be some orientation of molecules at intermediate angles, but the majority will line up as depicted.

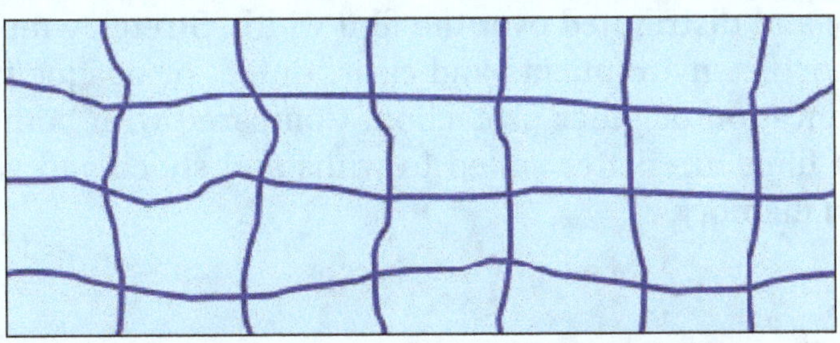

Polymer Molecules in Uni-axially Oriented Film.

Generally, orientation is imparted by blown-film extrusion process or on tenter frames. When the film is quenched after orientation, the molecules are "frozen" in the oriented position, but they still 'remember' their original shape or regain their memory to return to that shape if the film is reheated or exposed to the orienting temperature. When this does happen, obviously the film shrinks around the product.

Stretch Film

Stretch film is defined as stretchable, elastic, continuous thin plastic film, which is stretched and wrapped around one or more items to protect them from the environment or unitize for handling, storage or shipping. Stretch film is also used to some extent for bundling smaller units and lumber. They are also used for super market tray wraps.

In general, stretch film means a collective name for all types of plastic films, which are wrapped round a package under some form of mechanical stretching. It is a film, which can be cold-stretched in longitudinal and transverse directions without application of heat and which when stretched round a pack maintains a tension for a long period. A strong degree of transverse orientation gives stretch film a good extensibility in longitudinal direction and an increased total strength.

Stretch films entered the market in the early 1970s as a replacement for shrink wrap, used to unitize non-returnable glass bottles, pet jars in a tray, etc. The most common type of stretch film is the 'cling' type which is easy to use since the wrapping is completed by cutting the film between the load and the film roll and merely wiping the loose tail of film against the load. The film to film adhesion i.e. cling, holds the tail in place. Other less common way of attaching the film tail is adhesives, heat sealing, mechanical fasteners and tying.

Advantages and Disadvantages of Stretch Wrap

Advantages

Compared with shrink film and bags, stretch wrapping affords large energy saving and does not require the availability of fuel. Compared with shrink bags in particular, it simplifies inventory. Compared with strapping, stretch wrapping eliminates the need for corner boards and prevents cutting or crushing of the load, since the load-holding force is adjustable and distributed over the film width. Stretch wrapping offers better protection from hostile environments and clear, tinted, or opaque films are available for product identification or pilfer protection. Compared with both shrink films and strapping, stretch films are better suited to withstand shock and vibration owing to their elasticity and memory.

Disadvantages

Stretch wraps generally have less moisture resistance than shrink bags, but a top sheet can be dropped over the top of the load for added protection before stretch wrapping. The cling property that makes stretch wrapping possible, can also promote load-to-load sticking and abrasion. Stretch films cannot be used to compress a pallet load in the vertical direction since it has its primary holding force in the direction of wrap, i.e. horizontal.

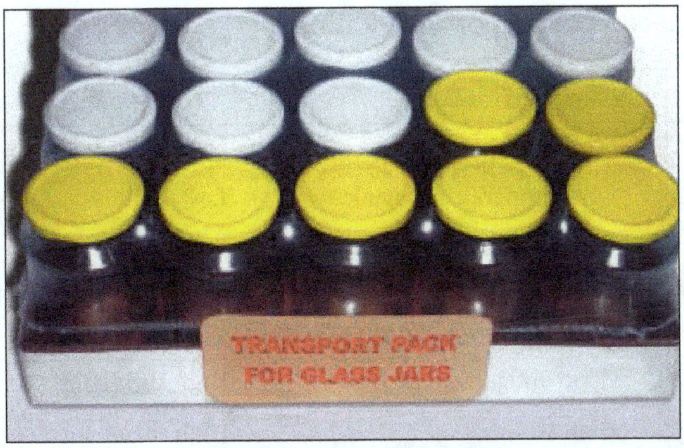

Shrink Pack of Glass Jars.

Important Material Properties of Shrink/Stretch Film

Low Melt Flow Index (MFI) for PE film (e.g. 0.1 to 0.6) indicates high melt viscosity, which in turn depends upon the fact that the material is composed mainly of long-chain molecules i.e. has a higher molecular weight. Therefore, low MFI of LDPE gives a logical guarantee for high film strength, often a prime requirement for wrapping goods or pallet loads. As MFI has direct relevance to orientation, stretch properties, strength, etc., shrink/stretch films are divided into 3 MFI classes:

- MFI < 0.6: For heavy goods and pallet loads.
- 0.6 < MFI < 1.0: For lighter packages where high transparency is valued more than the maximum strength.
- MFI > 1.0: For applications, where strength is of minor importance; used for textiles and such products where a weak shrinkage force is desirable.

Strawberry Punnet - Stretch Wrapped.

Table gives the comparison of performance of various shrink films. The typical properties of the shrink films are given in Table.

Table: Performance comparison of shrink films.

Film Type	Advantage	Possible Problems
Polyethylene (low density)	• Strong heat seals. • Low temperature shrink. • Medium shrink force for broad application. • Lowest cost.	• Narrow shrink temperature range. • Low stiffness. • Poor optical property. • Sealing wire contamination.
Polypropylene	• Good optical appearance. • High stiffness. • High shrink force. • No heat sealing fumes. • Good durability.	• High shrink temperature. • High shrink force, not suitable for delicate or fragile product. • Brittle seals. • High sealing temperature.
Co-polymers	• Strong heat seals. • Good optical appearance. • High shrink force. • No heat sealing vapours.	• High shrink force, not suitable. • For fragile products. • Higher shrink temperature. • Higher heat seal temperature. • Lower film slip-may give machine problems.
Poly Vinyl Chloride	• Lowest shrink temperature. • Wide shrink temperature range. • Excellent optical appearance. • Controlled stiffness by plasticizer content control. • Lowest shrink force for wrapping fragile products.	• Weakest heat seals. • Least durable after plasticizer loss. • Toxic and corrosive gas emission from heat sealing, therefore good ventilation required. • Durability problem at low temperature. • Low shrink force inhibits use as a multiple –unit bundling film. • Low film slip causes machine wrapping difficulties.
Multilayer Co-extrusion	• Excellent optical appearance. • Good machineability. • Low shrink temperature.	• In co-extruded films, one ply compensates for the deficiencies of the other. As a result, they are superior films with no significant performance shortcomings. • The wide variability in layer composition and number of layer makes performance analysis difficult.

Table: Typical shrink film properties.

Film Type	Tensile Strength psi (mpa)	Elongation (%)	Tear Strength gf/mil (m N/m)	Maximum shrink (%)	Shrink Tension psi (mPa)	Film shrink temperature range °F(°C)
Polyethylene (low density)	9000 (62)	120	8 (3.1)	80	250 - 400 (1.7 - 2.8)	150 - 250 (65 - 120)
Polyethylene (low density irradiated)	8000-13000 (55-90)	115	5-10 (1.9 - 3.9)	80	400 (2.8)	170 - 250 (75 - 120)
Polyethylene (copolymer)	19000 (131)	130	7 (2.7)	50	450 (3.1)	180 - 260 (85 -125)
Polypropylene	26000 (179)	50-100	5 (1.9)	80	600 (4.1)	250-330 (120-165)
Polyester	30000 (270)	130	10-60 (3.9 - 23.2)	55	700 - 1500 (4.8 - 10.3)	170 - 300 (75 - 150)
Poly Vinyl Chloride	9000-14000 (62-97)	140	Variable	60	150-300 (1 - 2.1)	150-300 (65 - 150)

Advantages of Polyolefin Film

Strength and Safety of Packs

- Durable and tear-resistant with tough seals.
- Neither easily embrittles during packing and delivery nor discolours with ageing.
- Strong and flexible packing that ensures tight pack even at refrigerator and freezer temperatures.
- Excellent wrap around multipacks and uneven or irregularly shaped product.
- Perfects shapes due to high shrinkage.
- Ensures consistent and uniform seals.

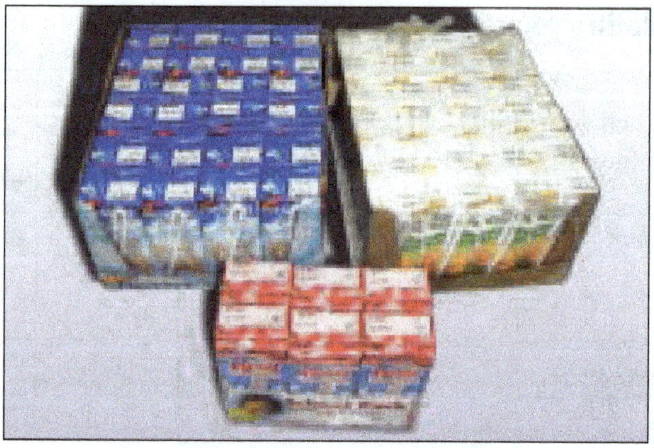

Shrink Wrapped Cartons with Tray.

High Shelf Appeal

- High gloss and clarity of added pack value and visual appeal.
- Provides tamper evidence and tamper resistance.
- Excellent printability for added appeal.

Increased Profit and Productivity

- Reduces packaging cost by as much as 50% or more.
- Cheaper than other substitutes like carton and packaging (wrapping) paper.
- Absence of acids, which corrode machine and sealing wire, thus saving on maintenance costs.
- Ease of processing on manual, semi-automatic or automatic shrink-wrapping machine.

Environment Friendly and Safety to User

- Polyolefin films are free from toxic and odour during processing.
- Approved for direct food contact by the USDA and FDA in United State of America, Canada and Europe.

Table: Comparison of shrink wrapping and stretch wrapping.

Shrink Wrapping	Stretch Wrapping
Equipment (heat shrink tunnel) is more expensive.	Equipment (relatively simpler and smaller) is generally less expensive.
Sealing required, particularly for overwrapping	Normally, no sealing is required.

More energy required: Orientation plus heat energy to shrink.	Less total energy - only stretching, but no heat.
May not be suitable for heat sensitive items.	Can even be used under cold conditions (refrigeration).
May distort under transit conditions.	Generally, retains load more tightly.
Wrapped items (two or more) may stick together.	Virtually, no film-to-film sticking.
Stick film takes up uneven contours more readily.	May create areas of higher tension due to irregular products.
Needs different film widths for a range of sizes.	Needs fewer reel widths for a range of sizes.
Can be printed using distortion printing where shrinkage is uniform and well controlled.	Easier to print as stretch is mainly in one direction.
Can provide better weather protection, particularly with a total over-wrap. Adds to general climatic protection.	Less protective, may not be totally waterproof. Lower climatic protection.
Uses generally more film (heavier gauges).	Uses less film.

Differences in shrink and stretch wrap for tray packaging are compared in table below. Similarly shrink bags and shrink rolls for tray packaging are compared in table.

Table: Comparison of stretch wrap and shrink wrap for tray packaging.

Parameter	Stretch Wrap	Shrink Wrap
Process	Not easy to pack tray by stretch wrapping.	It is easy to pack tray by shrink wrapping.
Cost	Slight higher than shrink wrap.	Less than stretch and CFB box packaging.
Machines	No machine for the tray packing.	Semi Automatic or Automatic L-sealer with shrink tunnel.
Time Required	More time required.	Less time required than other process.
Manpower	More manpower required.	Less manpower required.
Production	More production as compared with the other.	Less production as compared with the other.

Table: Comparison of Shrink Bag (Pouch) and Shrink Roll for Tray Packaging.

Parameter	Shrink Bag Pouch	Shrink Roll
Speed of process	Slow speed of operation.	Speed of operation is very fast.
Steps involved	Steps involved in operation are more.	Steps involved in operation are less.
Cost of packing	Cost of packing is more.	Cost of packing is less.
Material required	More material will be used.	Less material will be used.
Finish of packing	Finish is not uniform.	Finish will be uniform and much better.
Manpower required	More manpower required.	Less manpower required.

Shrink and Stretch Sleeving

During the past decade, the growth of sleeving has been quite significant. Earlier, sleeving was mostly used for cosmetics, toiletries and personal care product applications. Currently, however it is used for many food packaging applications, including drinking water and other drinks. Plastic sleeves are basically of two types:

- Stretch sleeves.
- Shrink sleeves.

Stretch sleeve labels are usually manufactured from LDPE, PP and PVC in a tubular form and then flattened and rolled. Depending upon the type of application, machine used, the sleeve manufacturer may need to incorporate perforations between the adjacent sleeves, which helps to separate from one another in order to form the individual sleeves, immediately prior to application to the containers. This separating action takes place as a result of differential speeds within the sleeve transfer station.

The specially designed stretching mandrels open up the sleeves before they are transferred down onto the container. Then the stretch sleeves conform to the contours of the containers. From the marketing perspective, it is advisable to have cut-off knives as an alternate to perforation, which may result in improper upper and lower edges of the sleeve.

A typical film thickness for stretch sleeves is 55-70μ and application speeds up to 800 containers/ min. are possible, depending upon the type of machine. As, unlike shrink sleeves, no heat is used, stretch sleeves can be used for applications for a diverse range of plastic containers. This technique is particularly suitable for chilled products like fruit juices where the container is likely to change shape during filling or subsequent storage. Stretch sleeves are generally used on round containers, where the graphics do not require any form of orientation with regard to the geometry of the containers. However, it is possible to orientate the print as is necessary for square or other non-round plastic containers.

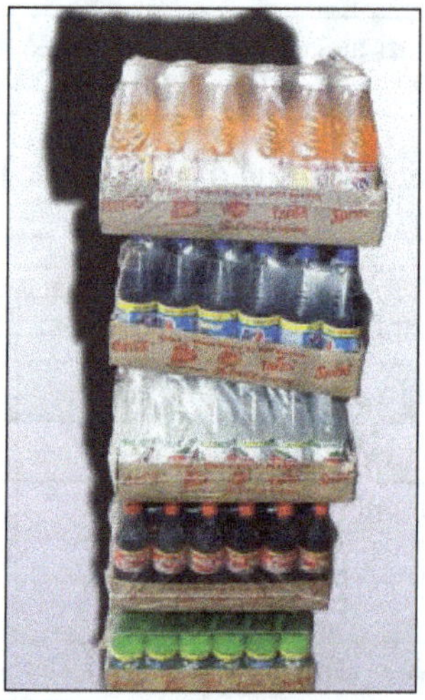

Shrink Wrapped Transport Pack for Carbonated Soft Drinks.

The method of manufacture, reel feed and application for shrink sleeves are almost similar in many respects to those for stretch sleeves. The main difference is, of course, the fact that heat is necessary in order to create the required degree of shrink to ensure that the sleeve conforms totally to the profile of the containers being labelled, without any evidence or wrinkling or distortion of the graphics. If the plastic film needs to be shrunk by varying degrees in different areas of the container profile, then the original design of the graphics need to take into account the subsequent shrinking operation.

One of the biggest advantage of shrink sleeve labels is their ability to provide 360° graphics around very unusual pack shapes. In addition, shrink-sleeves can serve a dual role, firstly being the products label and secondly providing a seal or tamper evident device. In such cases, the sleeves are designed to encompass all or most of the body area of the containers and in addition, to particularly encapsulate closure area also. The most common material for shrink sleeves is PVC, but OPP, PET and oriented polystyrene sleeves can also be used.

Heat Sealers

Heat sealers are appliances with a heating element that uses an electric current to heat and seal thermoplastic bags to form a sealed unit. The heating bar is pushed down on the thermoplastic material to melt it together and it is during the cooling stage that the seal is formed. Heat sealers can also fuse aluminium foil. Heat sealers come in different sizes and methods of application. The type of heat sealer you need depends on the product, seal strength and integrity of the seal you require.

Hand sealers are operated by pushing the sealing bar down by hand. As the bar is lifted the heat switches off to allow the seal to cool and firm up. A foot sealer is a heat sealer on a stand and uses a foot pedal to push the heating bars down. Hood shrink machines apply heat and shrink the plastic bag around the contents in a heat chamber. Gas heat guns apply heat to shrink a tight thermoplastic layer around an object. You can choose from a range of heat sealers, with or without a cutting apparatus, and either hand or foot operated. A cutting apparatus is a sliding blade that cuts off excess packaging and is useful for cutting poly tubing to form an individual bagged unit.

Heat sealers are used by different industries. They are an excellent application for the food industry. Keeping packaged food airtight and fresh, prolonging the shelf life. The pharmaceutical and medical industries also use heat sealing equipment to keep dressings, medication and surgical implements sterile. Retail stores such as newsagents, lolly shops or small home-based businesses use heat sealers to protect goods from dust, moisture and han-dling.

Impulse Heat Sealers

Impulse heat sealers only apply heat when the heating jaws are pushed down onto the area to be sealed and the set time is reached. This type of heat sealing is usually used for fusing thermoplastic materials that don't need high temperatures to form a seal. Impulse heat seals multi-layered thermoplastic and foil tubing to form a moisture barrier and airtight pouches. These are the safest heat sealers as the element is only hot when the heating jaws are closed. Burns are more likely to be avoided. Impulse heat sealers are more economical to use as the power usage is intermittent and the sealing jaws are heated instantly, rather than requiring time to heat up.

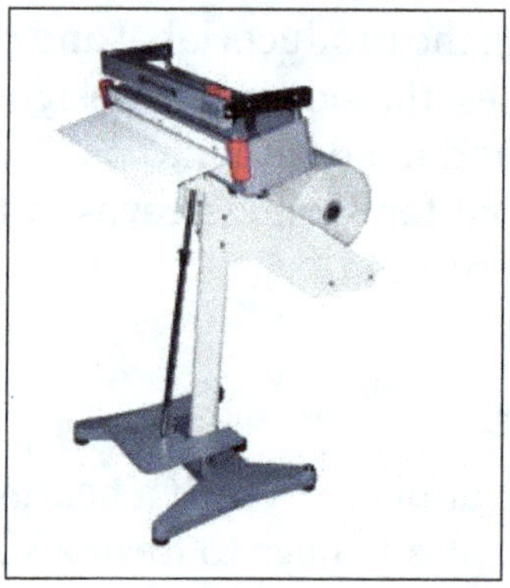

Constant Heat Sealers

Constant heat sealers provide a high continuous heat in the sealing jaws. The temperature is more controlled in these types of heat sealers. Thermoplastic or poly-coated materials that are suited for constant heat need to be thicker and able to handle high temperatures without becoming weak or melting.

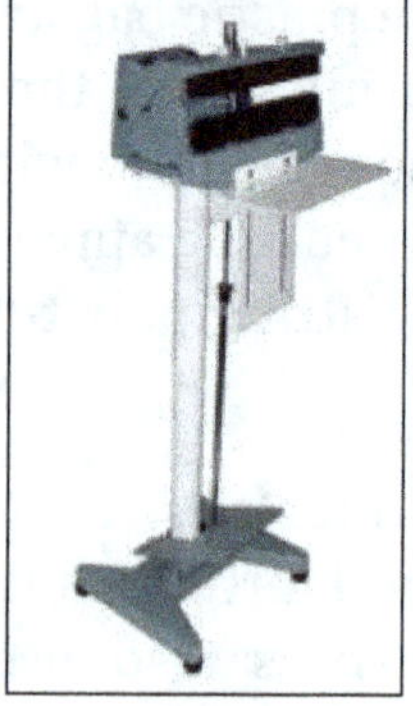

Constant Heat Roller Sealers

Seals Polyethylene or Cellophane by rolling the Teflon coated wheel along the material to be sealed. These portable sealers can make a round oval or square textured seal and are easy to use. Has an adjustable temperature range from 30-220 degrees. Is ideal for sealing plastic bags.

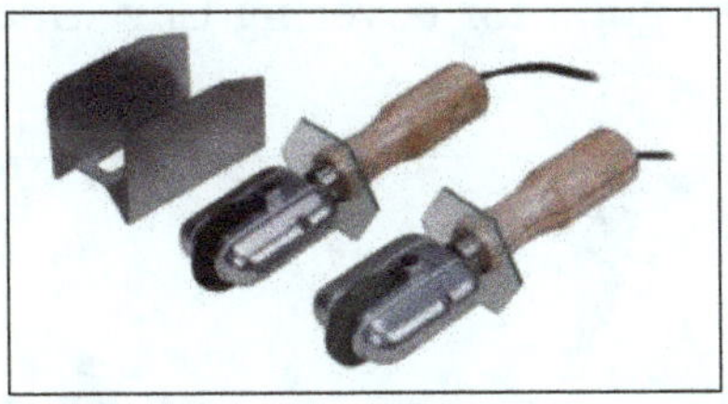

Band Sealers

Band Sealers come with a conveyor belt for a faster packing operation. These machines often have accessories such as an alphanumerical embossing wheel. Temperature controls let you set the heat levels and wider conveyor belts can be installed.

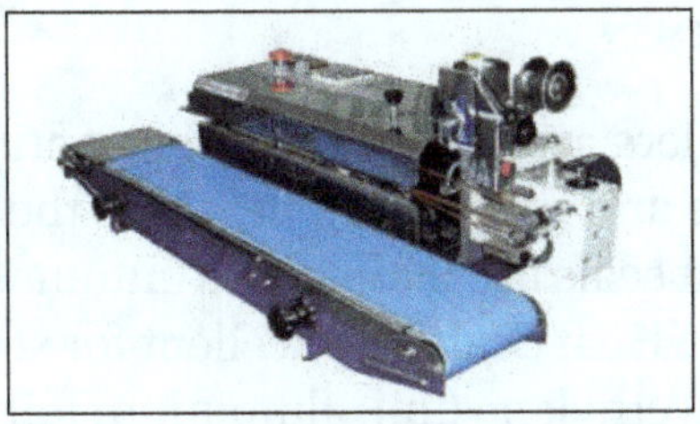

Hood Shrink Sealers

Hood Shrink Machines apply heat to shrink a plastic film around a product in one step. Hood shrink machines cover the item and then send it through a heat tunnel or oven to shrink the wrap. This process is often done by automated equipment, but semi-automatic and manual shrink wrap machines are also available. The shrink and cooling times are programmable according to the items and shrink material being used.

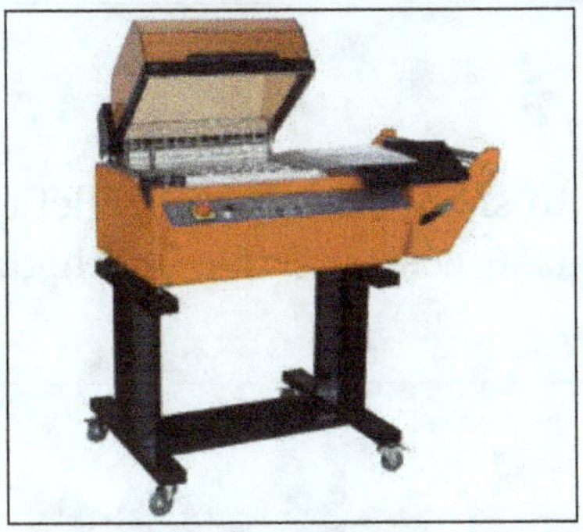

Shrink-a-pack

A heat machine that shrink wraps packages with the included heat gun and a film dispenser attached. This heat sealer is good for fresh produce and odd-shaped items such as hampers. For use on PVC, polypropylene or polyolefin shrink film. Great for small businesses and retailers such as chemists or newsagents who need to bundle items. This heat sealer is also good for covering Christmas hampers, perishables and non-perishables.

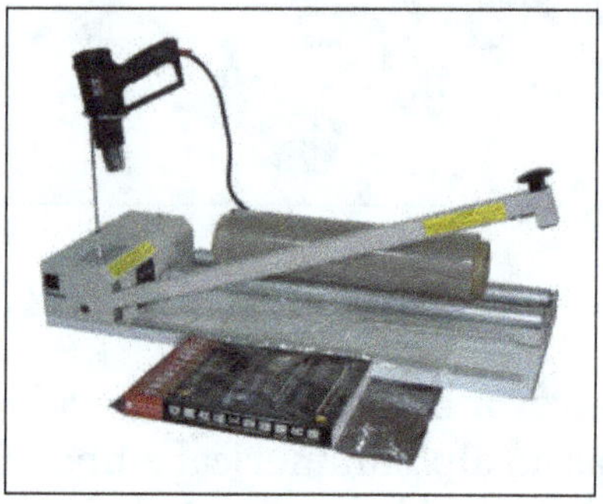

Automatic Heat Sealers

Automatic Impulse Heat Sealers are designed for the continuous sealing of thermoplastic films such as polyethylene and polypropylene. Once the correct control settings have been programmed, consistent seals are maintained automatically, making each seal consistent. A Constant Automatic Heat Sealer is excellent for sealing poly cello films, humidity-proof cellophane film, polyethylene film, aluminium foil coated bags and gusset bags.

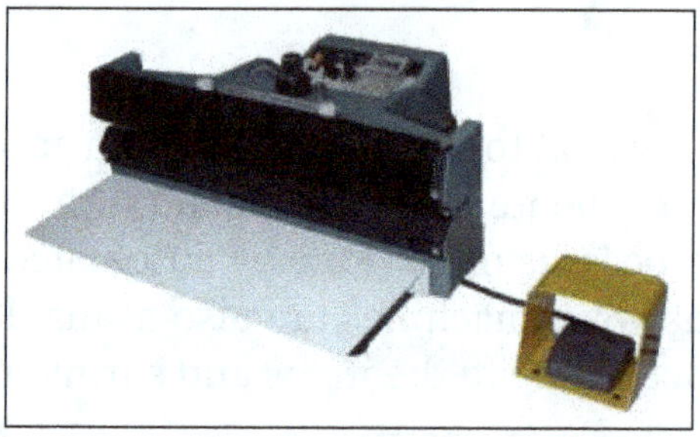

Gas Shrink Gun

Gas Shrink Guns are used with shrink polythene pallet covers, pallet bags, polythene sheeting, shrink films. It's a fast, cost-effective and highly protective method of dispatching goods.

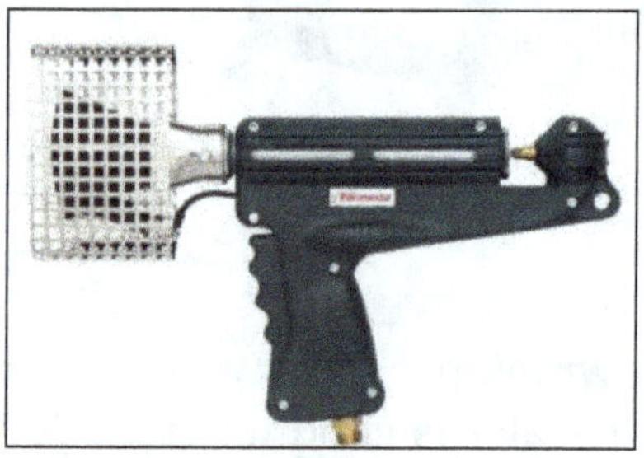

Seal Strength

Seal strength relies on the joining strength required for the object being packaged. Seal integrity relies on the level of minute breaks in the seal or tiny missed areas of seal that microbes can travel through and contaminate goods. Seal integrity is paramount for the medical and pharmaceutical industry. For proper sealing of goods, it is very important to factor in the materials, heat seal times and the requirements of individual items. The element in a heat sealer will eventually burn out. Luckily they are easy to replace with readily available spare parts and repair kits.

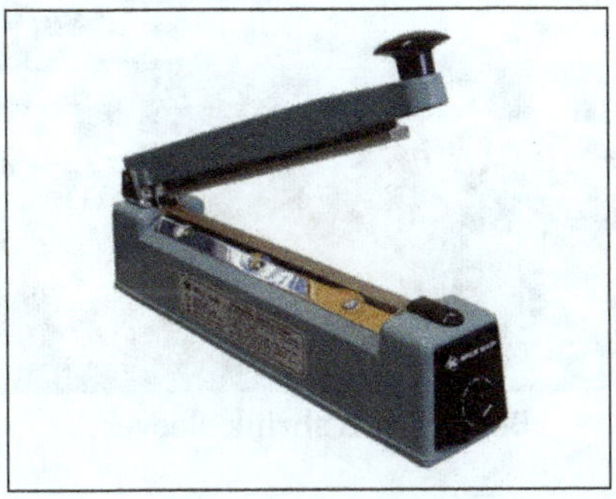

Shrink Tunnels

A shrink tunnel (also commonly known as a heat tunnel) is a metallic tunnel that sits above or is built around a conveyor. The tunnel has a heating element that heats the products as they move through the machine. Your products are conveyed through the tunnel and are cloaked with loosely fitting shrink film. When your products move past the heating elements, the film shrinks uniformly around the items as they are exposed to the heat.

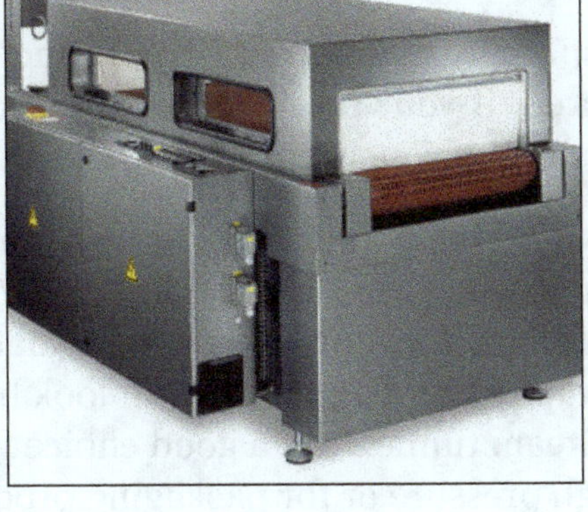

Different Kinds of Shrink Tunnels

Steam Shrink Tunnels

Steam shrink tunnels use steam rather than hot air to produce the heat required for shrinking plastic substrates on products. Steam tunnels are most commonly used for tamper-evident packaging applications, such as shrink sleeves. While steam shrink tunnels are not pneumatic machines, and therefore do not require an industrial level air compressor, they do require a water supply. As such, when considering purchasing a steam shrink tunnel, you need to understand that these machines will be more expensive because they require the use of a utility.

Bottles with shrink sleeves.

Self-contained Steam Shrink Tunnels

There is a variation of the steam shrink tunnel known as a self-contained steam shrink tunnel. The self-contained tunnel differs from the standard steam tunnel in that it has an internal steam generator which generates the steam without additional attachments. In contrast, the traditional steam tunnel needs steam piped into the machine which requires additional components. You may be required to have a steam engineer on the payroll to operate your steam tunnel as some states require special training and licenses to operate these types of shrink tunnels.

Steam shrink tunnels are specially made to run on product lines in which quality and high-end graphics are favored. If your products require spot on registration, brilliant colors, perfect fonts and long term consistency, steam shrink tunnels are for you.

Steam shrink tunnels allow for highly uniform heat transfers which are superior when compared to hot air systems. Steam-based machinery provides high precision registration capabilities and better shape forming. This means that the sleeves will fit tighter and more evenly and the graphics and aesthetics will look better on the sleeves resulting in better shelf appeal. Steam tunnels are a good choice when packaging containers which are packed under high pressure or for packaging products that are flammable.

Convective Shrink Tunnels

With convection shrink tunnels (also known as hot air shrink tunnels) the machines produce hot air which is blown onto a plastic sleeve that surrounds the item being packaged. The heat shrinks the sleeve around the product evenly. The hot air is blown from a module known as a "baffle." These modules direct the heat in any required direction which allows for superior distribution of heated or cooled air. The packaging of board games and frozen pizza boxes are common applications for this type of machine.

Radiant Shrink Tunnels

Radiant shrink tunnels use radiation as the tunnel's heat source. The sleeves in the machine absorb this infrared radiation. Radiant shrink tunnels are perfect for packaging applications that require lower shrink percentages. Radiant tunnels are commonly used for shrinking seal bands and other types of tamper-evident packaging materials.

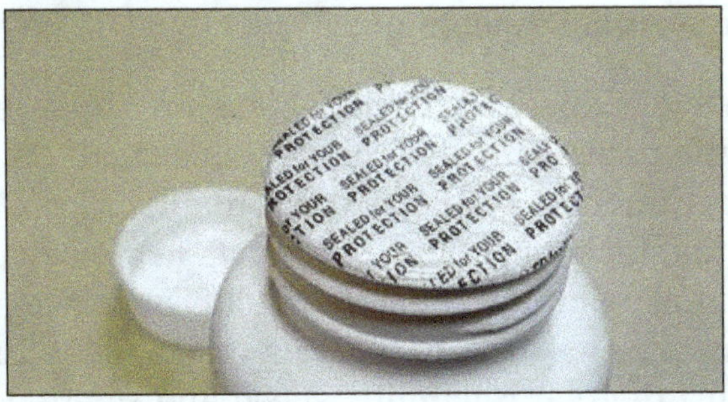

Palletizer Machine

The palletizer machine is a device that is used to stack factory goods and products on to a pallet. The first palletizer was built and installed in 1948. Row-forming palletizers were introduced in the early 1950's that were used to arrange the goods in a row. Inline palletizers were built in 1970's when high speed was increasingly needed for palletizing. Robotic palletizer machines, which have a single arm, were introduced in the early 1980's.

A robotic palletizer machine is a complex machine which is used to pack boxes at a very high speed. Whenever you talk with anyone about a palletizer machine, the first thing which will come to their mind is an automatic or robotic palletizer machine. There must be a means to bring products and a pallet together. This usually requires two or more rolling platforms, but in some cases workers move pallets manually. When the palletizer machine has the boxes and the pallet in reach, it picks up the boxes and neatly arranges them onto a pallet. The boxes are usually of the same size but some palletizer machines are programmed to recognize differently-sized packages. These palletizers are capable of distinguishing the different sizes and they put the boxes on their pallets. The palletizer machine neatly stacks the boxes into rows and packs (or wraps) the pallet. In some cases, pallets are moved away from the machine and packed/wrapped either by hand or by some other machine.

There are several types of palletizer machines such as high level in-feed, low level in-feed, semi- automatic palletizers and robotic palletizers. Semi-automatic palletizer machines are cheaper and are available at affordable prices. This type of palletizer machines is designed for stacking goods or products in a particular pattern. Semi-automatic palletizers require minimum operator input. Low level in-feed palletizer machines are fully automatic. This type of machine makes handling of boxes very easy and these machines ensure great reliability, speed and safety. High level in-feed palletizer machines are also fully automatic, can have a multi-product palletizer, and can serve multiple production lines.

Palletizing Machine

Conventional manual packing has always been in use, but over time, industrial development and the increase in amount of consumer items and industrial products led to a need for the technical world to develop a machine that would reduce human work load and, in the end, one that would not need a human to pack the stock. The palletizing machine was first introduced in 1948, has and since gone through considerable evolution. Now, palletizing machines are used in all kinds of industrial packing, ranging from small consumer products to high tech industrial products and goods. Gears, cartons, bails, bags or bundles can be handled with a palletizing machine.

Most of the palletizing machines available on the market are operated automatically and are extremely simple to use. Palletizing machines are available in two forms: the conventional palletizing machine and the robotic palletizing machine. A conventional palletizing machine has two major parts, a low level and high level part. The palletizing machine starts its work with the low level part rotating the pallets in a pattern, and then starting to stack the material in rows. The rows are built to a certain height, and finally the pallet is ready to be handed over to the high level part. The High level palletizing machine works similar to the low level palletizing machine except that it works at a much higher speed than the low level palletizing machine, and pallets are moved up and down as required while the base layer remains stationary.

Advantages of Palletizing Machines

Advantages associated with a palletizing machine are numerous. Palletizing machine has worked in packing for years and have proven to be extremely efficient. The benefits associated with a palletizing machine provoked researchers to keep on improving the palletizing machines and today there are very complex palletizing machines being used by huge industries. The advantages are:

- Significant Space Saving: Palletizing machines require less space than human manpower on location and are available round the clock. Complex palletizing machines further reduce unnecessary manpower at a given work place.
- Ease of Operation: Palletizing machines are easy to use and are extremely simple in their operation.
- Fast and Gentle: Palletizing machines are designed to handle each object with gentle care and to assure that none of the objects and pallets that a palletizing machine handles is damaged.
- Cost effectiveness: Besides many other advantages, a palletizing machine is cost effective. Reducing manpower at a particular site would in the long run result in saving much more than what is spent to procure a palletizing machine.

Robotic Palletizer

The increasing spread of industrial robots has allowed the optimization of the palletizing process, enabling an increase in efficiency, productivity and versatility compared to the "classic" methods of palletizing (manually or with a Cartesian Palletizer). These features are further highlighted by the market appearance of the so-called "palletizing robot": particular types of robot with 4 axes (now offered by leading manufacturers worldwide such as ABB, Kuka, Fanuc, Comau, etc.) designed specifically for palletizing and which can offer very high performance with regard to speed and payload.

The technical leaps of this era have given palletizing robots a new dimension with diversity in their use and efficiency in their output. As a result, the robotic palletizer has provided the market with a new automated cost-justified solution with their capability to handle more than one section. Today, robotic palletizer systems are available at different prices varying with the level of robotics and with the number of accessories provided with the palletizing robot.

Usually, a robotic palletizer is integrated with software for generating and optimizing palletizing patterns: the user inputs the size of the pallet and of the items to be palletized, and the system automatically generates the palletizing program and sends it to the robotic palletizer. The advantages of this approach are many. In particular:

- Increase of productivity: Palletizing programs are created and sent to the palletizing robot in a short time without any specific requests for assistance or expertise from the staff.
- Reduction in time and cost of palletizing: The use of a robotic palletizer delivers high performance in terms of speed and versatility (the robotic palletizer can handle multiple input lines and/or storage zones and can automatically adapt itself to different product types).
- Reduction of transport costs: Optimizing the amount of product on the pallet can be translated into logistics and transportation cost savings estimated at well over 30%.

Automatic Palletizing

Automated palletizers are the machines which are used to place things on pallets in a particular pattern. Automatic palletizing machines come in different types to handle different types of automation requirements and types of packaging, ranging from bags to drums, from crates to cases, etc. There are two types of automatic palletizing machines: fully automated palletizers and semi-automated palletizers. Fully automated palletizers do not need any type of instruction from the operator and can perform their function without any intervention.

Another classification of the automatic palletizing machines can be made between low level, high level and robotic. Typically, robotic palletizers can arrange products, goods, boxes, crates, drums, etc. on a pallet in a tier made by the operator of the automatic palletizing machine with the help of so-called Pallet Loading Software.

Automated palletizers are mostly used where packaging speed is important and where the product being shipped is fragile and/or heavy: an automated palletizer applies less stress on the products or goods while putting them on a pallet.

Advantages of Automated Palletizer

There are so many benefits of automatic palletizing machines:

- Automatic palletizing machines can significantly increase the speed and the efficiency of the line.
- Automatic palletizing machines increase flexibility and throughput of the line.
- Automatic palletizing machines can increase load accuracy and can place goods or products more tightly compared to when they are placed manually.
- Automatic palletizing machines can reduce the amount of products damaged during handling.
- An automated palletizer can reduce labor injuries associated with manual handling.
- An automated palletizer is cost effective.

Palletizing Systems

The machines that are used to place goods or different products on a pallet are called palletizers and the whole system is called the palletizing system. There are different kinds of palletizing systems such as robotic palletizing systems, high level palletizing systems, and low level palletizing systems. The main advantage of a palletizing system is that it allows objects like goods, drums, crates and cases to be placed on a pallet in the most efficient way possible: therefore a palletizing system helps to increase the plant's production.

A robotic palletizing system is inexpensive and available at affordable prices. This type of palletizing system requires minimal operator input and doesn't require any type of manual lifting or handling. Most recent palletizing systems are *fully automatic* and ensure safety, speed and great reliability and, last but not least, have great throughput and are capable of serving different production lines with different types of products: automatic palletizing systems can automatically recognize objects of different shapes and sizes and can serve several lines at the same time.

Advantages of Palletizing Systems

- Flexibility of Controls: Compared to manual palletizing, a *robotic palletizing system* provides the user with a model that gives them the flexibility to handle any kind of product (even if it is very heavy or large).
- Safety: The in-built safety features of a modern automatic palletizing system offer greater safety compared to manual palletizing, thereby reducing labor injuries.
- Speed and efficiency: Automatic palletizing systems improve the efficiency and the speed of the plant. A robotic system can perform better and palletize a hundred times faster than a human.

Palletizing Equipment

No palletizing operation can be considered fully automatic without the integration of palletizing equipment such as: pallet conveyors, pallet lifts, pallet dispensers and stackers, shuttle cars, etc. Typically, palletizing equipment incorporates some combination of mechanical, pneumatic and servo functions applied to meet the specific needs of palletizing. Below, some of the palletizing equipment that is most commonly used in palletizing and end-of-line packaging will be briefly described:

- Pallet conveyors: There is palletizing equipment that makes it possible to automate the process of loading and unloading pallets from the palletizing system, in order to expedite and streamline the machine, eliminating time losses related to manual operations. Typically a pallet conveyor is built with rollers, if the pallet is moving parallel to the wide side; or with chains, if the pallet is moving parallel to the narrow side. This type of palletizing equipment is widely used in industry due to its robustness and simplicity and to the low cost of purchase and maintenance.

- Pallet stackers and pallet dispensers: Pallet stackers and pallet dispensers are devices that allow the accumulation and distribution of pallets automatically as needed. This palletizing equipment is increasingly common in the industry, particularly in the case of highly automated palletization, because they allow the palletizing machine to be automatically supplied, avoiding the downtime associated with manually loading empty pallets.

- Pallet shuttle car: The pallet shuttle car is a piece of palletizing equipment that can be used as a means for taking a full pallet out of a palletizer at the same time that an empty pallet is supplied to the same palletizer. Similarly, with this palletizing equipment, the pallet can be moved automatically between the various palletizing stations, eliminating forklift traffic between the various ends of line packaging zones, palletizing stations and warehouses.

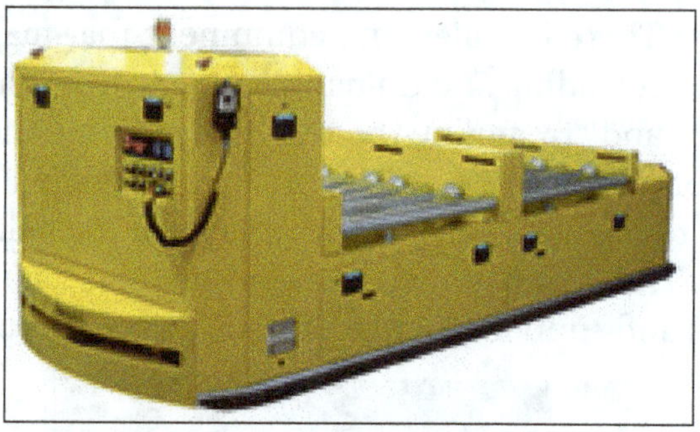

Mixed Palletizing

Mixed palletizing is the increasingly popular technique of placing products (typically boxes, bags or cylindrical containers) of different types and sizes on a single pallet. This need, which relates to the rising cost of transportation and logistics, requires optimizing space on pallets in order to place as many products as possible.

Mixed palletizing is a challenge that companies handle using increasingly complex and integrated palletizing systems. Typically, a mixed palletizing system consists of one or more lines of product delivery and one or more palletizing stations. The palletizing system (usually consisting of a robot) recognizes the products (through vision systems or other types of indexing), picks them up, and deposits them on the pallet in accordance with the assigned palletizing pattern.

Although deriving from traditional palletizing, mixed palletizing introduces a series of complications involving both mechanics and control logic. From the mechanical point of view, mixed palletizing requires special attention in the design of the gripper for picking the products: the device must in fact automatically adapt itself to various types of product in order to ensure a firm hold. From the control logic point of view, the major complications of mixed palletizing are linked to the need for automatically recognizing products and optimizating palletizing patterns. In particular, the problem of optimizating the layout of products on pallets, already difficult in conventional palletizing, becomes extremely complex in mixed palletizing due to the high number of possible solutions. It is therefore essential to use Pallet Loading Software for a mixed palletizing procedure that involves palletizing more than two or three different products.

Cartesian Palletizer

The cartesian palletizer is a type of automatic palletizer in which the movement of the picking and placing gripper is entrusted to a mechanical structure (portal) usually built with beams or structural forms in metal or aluminum. This structure, in conjunction with "axis-control" systems made with "drives" and/or "Plc", allows the gripper to palletize products by moving in the three directions of space (cartesian coordinates: x, y, z) and rotating around the vertical axis. Because of its construction type, the cartesian palletizer does not allow very high performance and, most importantly, severely limits flexibility and versatility in the face of changes in palletizing needs or layouts that can occur within a production facility. Nevertheless, the Cartesian Palletizer is still used, especially in not particularly "complex" palletizing systems, by virtue of its low cost of implementation and maintenance.

Bag Palletizer

The ever increasing need to automate processes in order to increase capacity and reduce production costs has resulted in an increase in demand also in the field of automatic bag palletizing. The bags are usually heavy and bulky, and their manual palletizing is therefore hard and exhausting: an automatic bag palletizer is therefore the best and cheapest solution for every company that needs to palletize bags and that wants to meet the market competitively. Modern bag palletizers are economical, flexible and provide a consistently well-packed pallet pattern that upgrades the line throughput and helps in reducing product damage associated with physical handling. Currently the two most common types of bag palletizers are the layering bag palletizer and the robotic bag palletizer. The main characteristics of these two types of bag palletizers are described below:

- Layering Bag Palletizer: This type of bag palletizer can process a wide range of products with very different sizes and weights. The solution used for the formation of the layer on the pallet, which is usually done row by row, enables high rates of work. Some models of layering bag palletizers can square and compress the layer just placed on the pallet to secure the shape of the bag and optimize the geometry of the pallet.
- Robotic Bag Palletizer: The robotic bag palletizer is a special kind of palletizer based on an anthropomorphic robot. The bag palletizing robot is equipped with a gripper designed according to the characteristics of the product and can lift and palletize bags weighing hundreds of pounds. Usually, the gripper also has special devices for lifting the empty pallets in order to fully automate the process of bag palletizing. The high flexibility of this type of bag palletizer and its competitive price mean that the robotic bag palletizer is finding increasing use in industry.

Bottle Palletizer

The need to palletize bottles (whether glass or plastic, full or empty) in order to reduce the costs of transport and logistics is growing in many sectors of industry, from food to glass. The manual palletizing solution is no longer acceptable both for its low efficiency and for its high labor cost. In the last decade, the use of the automatic bottle palletizers has spread, ever more widely.

These particular types of palletizers are built specifically for palletizing bottles and can achieve work rates unimaginable for manual palletizing. From the construction point of view, these palletizing machines are divided into layering bottle palletizers and robotic bottle palletizers. In the case of layering bottle palletizer, an entire layer of bottles is pre-formed on a special surface (preformation table) and then the layer of bottles is stacked on the pallet.

As for the robotic bottle palletizer, this is a robot (usually anthropomorphic) which is in charge of picking and placing bottles on the pallet. In this case the picking can be done either from the preformation table or directly from the bottle conveyor belt. Taking advantage of the high speed modern robots can work at and of the special grippers for picking the bottles (for example with devices for the quincequonce of rows), it is possible to eliminate pre-forming the layer of bottles, with countless advantages in terms of the performance and flexibility of the palletizing system. Usually, the gripper of a robotic bottle palletizer is also equipped with devices that allow the handling of pallets and packing materials (slipsheets, caps, etc.). The robotic bottle palletizer is also proposed as an alternative to traditional bottle palletizers in cases where the space in the plant is limited: a palletizer robot is less bulky than a traditional bottle palletizer.

Used Palletizer

Many industries which need to palletize their products are not able to run in the absence of a palletizer. The equipment has become an integral part of success for these industries and it is for this reason that the demand for new and used palletizers is increasing day by day. There is a huge difference between the amount paid for a new palletizer and for a used palletizer that has been reconditioned: you will pay nearly half the money this way and, to a great extent, support the growth of your business.

Buying a used palletizer is quite similar to buying a used vehicle: like used car dealers, many companies are specialized in buying and selling used palletizers of the best quality. It is possible to find a good range of used palletizers and select the one that satisfies your requirements. Important: when buying a used palletizer, check that it is covered sufficiently by a warranty. Usually, these companies provide a good percentage of coverage on various parts and offer technical support for the used palletizers. Some companies can also demonstrate their used palletizers for you.

Advantages of Used Palletizers

- A used palletizer is inexpensive compared to the new one and, at the same time, a *serviced* palletizer would work as efficiently as new equipment.
- The used palletizer could be repaired immediately by the service centers which did the sale. Spare parts for the used palletizers may be immediately available because they have been in the field for some years.
- A used palletizer would serve at the maximum capacity from the day of installation, while new equipment would not be good enough to produce the maximum quantity in the beginning.

Depalletizer

Depalletizers are robotic machines that unload (depalletize) packages or products from the pallet. The depalletizer allows the transfer of products and materials from pallets without the use of human personnel, saving thus time, money and effort. There are several types of depalletizers, depending on the features offered, but especially depending on the type of material and on the product to unload from the pallet. The lifting of products to be depalletized is usually done with one of three proven methods, i.e. the pneumatic system, the vacuum chamber, and mechanical fastening: depalletizing thus can adapt to any type of product.

Modern depalletizers are equipped with a control system that stores product data and the assignment of parameters such as speed, acceleration, clamping force (or the amount of depression when lifting with a vacuum chamber), in order to ensure dynamic, secure and careful product handling.

Usually, depalletizers are designed so they can take a whole layer of the pallet: this provides very high rates of work. But there are also special types of depalletizers, with appropriate vision systems, which can recognize and lift individual items from the pallet; this approach allows greater flexibility and the machine can handle mixed pallets: in this case the individual products can be depalletized and stored in different areas. The major depalletizer producers are able to provide depalletizing systems according to customer needs, to provide personnel to accompany the customers in the first depalletizing cycles, and to ensure an efficient and durable maintenance service.

Palletizer Manufacturers

The present internet world permits any palletizers manufacturer to avail machinery from any corner of the world. The countries strict with the export import policy also permit the local companies to import machines from the palletizers manufacturer. The palletizers manufacturer must have to work hard to produce their machines. They are not ordinary machines. They got to be checked enough before the delivery. The palletizers manufacturer would be available in Germany and Columbia. The other country palletizers manufacturer would demand more money. Same time the local country could not have palletizers manufacturer in their country due to many problems. In addition to it, huge investment is required. The palletizers manufacturer needs more staffs and more equipment to make the machinery. The palletizers manufacturer would have fixed orders. They work only based on their orders. If they get more orders, the palletizers manufacturer would take more time to deliver the machinery. For an example, if any company orders now it would be delivered even after one year.

The palletizers manufacturer is required bulk quantity of raw materials. Apart from that, company requires surplus staffs to manufacture the machinery. The palletizers manufacturer would be with a long term projects. The new company must have to wait to get the machinery from the palletizers manufacturer. The palletizers manufacturer also may not be interested in the new order. The reason is the palletizers manufacturer is already committed to supply his machineries. The palletizers manufacturer would be interested in upgrading the models. For that reason, the palletizers manufacturer would be interested in hiring design engineers. The design engineers working with the palletizers manufacturer would be paid enough money. That makes the design engineers not to switch over to new palletizers manufacturer. It is hard to be the palletizers manufacturer. The reason is the palletizers manufacturer must have everyday production. The palletizers manufacturer must have to wait for a long time to get a finished product form the factory. The finished product from the palletizers manufacturer must have to be tested before the delivery. The palletizers manufacturer must not get any remarks from a finished machinery. If he gets any negative remarks the palletizers manufacturer must be ready to rectify the mistake. The palletizers manufacturer must be ready to clear all due amounts. The palletizers manufacturer must have to pay huge money for his loan amount. The palletizers manufacturer service is an essential service.

Pallet Loading Software

One of the problems that occur in automatic palletizing is how to determine the optimum product placement on pallets. The greater the number of products theoretically placeable on a layer of the pallet, the more difficult it is to determine the palletizing pattern which allows the largest number of products to be palletized. This difficulty increases further when searching for solutions that guarantee a certain stability of the pallet. In the past, companies were limited to resort to "approximate" palletizing patterns based on manual calculations or by trial-and-error; today, fortunately, are available special softwares called "Pallet Loading Software" that cand find the optimal palletizing pattern in a very short time.

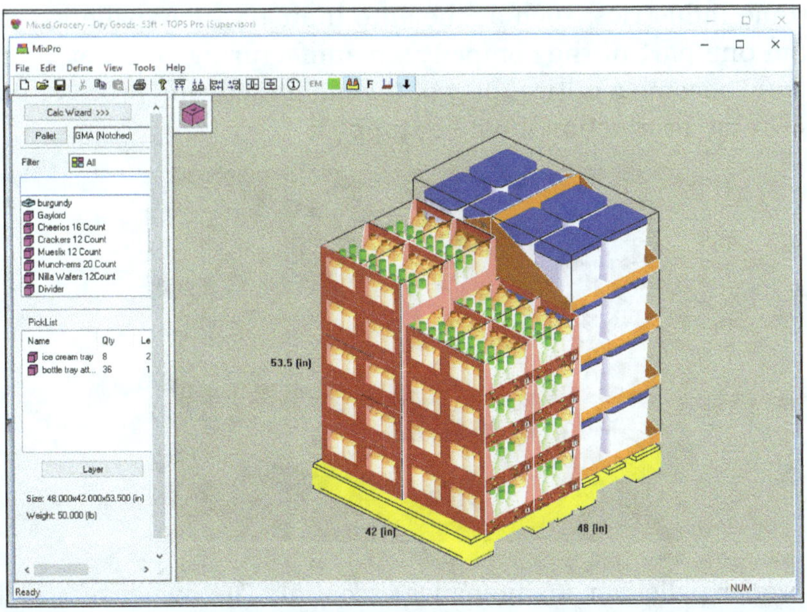

Some of the best Pallet Loading Software available is BonOnPallet: a software that has an extremely "user-friendly" interface and a sophisticated optimization algorithm that provides and displays the optimal palletizing pattern in a few seconds. An important feature of this pallet loading software is the ability to send data relating to the selected palletizing pattern directly to a robotic palletizer.

What is Palletizing?

Whether you spell it palletising or palletizing, this is simply the process of putting items onto a pallet for convenient, high-volume shipping. Palletizing has developed alongside the development of the forklift and the development of industry and globalization. Palletizing originally involved manual labor, putting boxes or heavy bags onto pallets for loading into trucks. The growth of palletizing has been made possible by ever more sophisticated palletizing systems, beginning with the palletizing machine installed by the Lamson Corporation in 1948. In the early 1950's, the row-forming machine was developed. This palletizing machine aids in the process of forming neat rows of product on the pallet.

The development of computers and robotics had clear implications for palletizing, as more and more of the work, and more and more of the control, could be handled by the palletizing machine itself. In addition, the variety of products that palletizers could handle grew to include items, such as bottles, which would be impossible or at least cost-prohibitive to palletize by hand.

Faster, more secure palletizing translates to more efficient logistics and lower transport risk for the company. As a result, palletizing units can be found in nearly every industry. Not all palletizing units are used for handling products destined for customers; some industries, such as the auto industry, may palletize parts for easy transport from one part of the company's manufacturing complex to another. Palletizing has truly become a critical part of the manufacturing process; if a palletizing process goes wrong, production must be stopped.

This makes proper palletizer maintenance essential. The owner of a palletizing unit may choose to have the unit serviced by the palletizer manufacturer, or may choose to send technicians to the manufacturer for training. Operation manuals for palletizing machines are nonetheless as complete as possible, as palletizer manufacturers know just how critical their products are.

Labeling of Food Packages

The objective of foodstuff labelling is to guarantee that consumers have access to complete information on the content and composition of products, in order to protect their health and their interests. Other information may provide details on a particular aspect of the product, such as its origin or production method. Some foodstuffs, such as genetically modified organisms, allergenic foods, foods intended for infants or even various beverages, are also subject to specific regulations.

Labelling of certain non-food products must also contain particular information, in order to guarantee their safe use and allow consumers to exercise real choice. In addition, the packaging of foodstuffs must adhere to production criteria in order to avoid contaminating food products. The nature of the labelling, the presentation of the foodstuff, and the publicity for it must not:

- Mislead the consumer as to the characteristics, the properties or the effects of the foodstuff.
- Attribute to any food the property of preventing, treating or curing a human disease (with the exception of natural mineral waters and foodstuffs for particular nutritional uses for which specific provisions exist).

Information on foodstuffs must be precise, clear and easy to understand by the consumer.

Responsibilities of the Manufacturer

The manufacturer under whose name or business name the food is marketed or the importer is responsible for the information relating to the foodstuff. They must ensure the presence and accuracy of the information in accordance with the applicable food legislation and requirements of relevant national provisions.

Mandatory Particulars

The mandatory particulars must be easy to understand and visible, clearly legible and, where appropriate, indelible. The height of the characters must be at least 1.2 mm (except for small-sized packaging or containers). The mandatory particulars concern:

- The name.
- The list of ingredients which are listed in descending order of weight and designated by their specific name.
- The substances causing allergies or intolerances (nuts, milk, mustard, fish, grains containing gluten, etc.).
- The quantity of certain ingredients or categories of ingredients.
- The net quantity of the food.
- The date of minimum durability or the 'use by' date.
- Any special storage conditions and/or conditions of use.
- The name or business name and address of the food business operator or importer.
- The country of origin or place of provenance for certain types of meat, milk or where failure to indicate this might mislead the consumer.
- Instructions for use where it would be difficult to make appropriate use of the food in the absence of such instructions.
- For beverages containing more than 1.2 % by volume of alcohol, the actual alcoholic strength by volume.
- Nutritional declaration.

The mandatory particulars concerning the name, net quantity and the alcoholic strength by volume shall appear in the same field of vision. The mandatory particulars must be in language which is easy to understand by the consumer, and, if required, in several languages.

Omission of Certain Mandatory Particulars

Specific provisions are provided for:

- Glass bottles intended for re-use.
- Small-sized packaging.
- The nutritional labelling of foodstuffs.
- Beverages containing more than 1.2 % by volume of alcohol.

Voluntary Information

Information provided voluntarily must meet the following requirements:

- It shall not mislead the consumer.
- It shall not be ambiguous or misleading.
- It shall, where appropriate, be based on the relevant scientific data.

Furthermore, the voluntary food information shall not be displayed to the detriment of the space available for mandatory food information.

General Labelling Requirements

Vegetarian/Non-vegetarian Symbol

A declaration to this effect shall be made by a symbol and colour code so stipulated for this purpose to indicate that the product is Non-Vegetarian Food. The symbol shall consist of a brown colour filled circle having a diameter not less than the minimum size specified in the table.

A declaration to this effect shall be made by a symbol and colour code so stipulated for this purpose to indicate that the product is Vegetarian Food. The symbol shall consist of a green colour filled circle, having a diameter not less than the minimum size specified in the table.

Table: Area and diameter of vegetarian/non-vegetarian symbol to be printed on package.

S. No.	Area of principal display panel	Minimum size of diameter, mm
1	Upto 100 cm^2	3
2	Above 100 cm^2 upto 500 cm^2	4
3	Above 500 cm^2 upto 2500 cm^2	6
4	Above 2500 cm^2	8

The symbol shall be prominently displayed:

- On the package having contrast background on principal display panel.
- Just close in proximity to the name or brand name of the product.
- On the labels, containers, pamphlets, leaflets, advertisements in any media.

Any article of food contains egg only as Non-Vegetarian ingredient, the manufacturer, or packer or seller may give declaration to this effect in addition to the said symbol.

Flavour and Colour

In case both colour and flavour are used in the product, one of the following combined statements in capital letters shall be displayed just beneath the list of ingredients on the label attached to any package of food, so coloured and flavoured, namely:

CONTAINS PERMITTED NATURAL COLOUR (S) AND ADDED FLAVOUR(S)

OR

CONTAINS PERMITTED SYNTHETIC FOOD COLOUR(S) AND ADDED FLAVOUR(S)

OR

CONTAINS PERMITTED NATURAL AND SYNTHETIC FOOD COLOUR(S) AND ADDED FLAVOUR (S)

OR

CONTAINS PERMITTED NATURAL*/AND* SYNTHETIC* COLOURS AND ADDED FLAVOURS

Food Additives

Food additives falling in the respective classes and appearing in lists of food additives permitted for use in foods generally, the following class titles shall be used together with the specific names or recognized international numerical identifications: Acidity Regulator, Acids, Anticaking Agent, Antifoaming Agent, Antioxidant, Bulking Agent, Colour, Colour Retention Agent, Emulsifier, Emulsifying Salt, Firming Agent, Flour Treatment Agent, Flavour Enhancer, Foaming Agent, Gelling Agent, Glazing Agent, Humectants, Preservative, Propellant, Raising Agent, Stabilizer, Sweetener, Thickener.

Details of Manufacturer

- The name and complete address of the manufacturer and the manufacturing unit, if these are located at different places and in case the manufacturer is not the packer or bottler, the name and complete address of the packing or bottling unit.

- Where an article of food is manufactured or packed or bottled by a person or a company under the written authority of some other manufacturer or company, under this or its brand name, the label shall carry the name and complete address of the manufacturing or packing or bottling unit as the case may be, and also the name and complete address of the manufacturer or the company, for and on whose behalf it is manufactured or packed or bottled.
- Where an article of food is imported into India, the package of food shall also carry the name and complete address of the importer.

Details of Quantity

The net weight or number or measure of volume of contents as the circumstances may require, except in the case of biscuits, breads, confectionery and sweets where the weight may be expressed in terms of either average net weight or minimum net weight.

Where a package contains a large number of small items of confectionery, each of which is separately wrapped and it is not reasonably practicable to exclude from the net weight of the commodity, the weight of such immediate wrappers of all the items of the confectionery contained in the package, the net weight declared on the package, containing such confectionery or on the label thereof may include the weight of such immediate wrapper if, and only if the total weight of such immediate wrapper does not exceed:

- 8% where such immediate wrapper is a waxed paper or any other paper with wax or aluminium foil under strip.
- 6% in the case of any other paper, of the total net weight of all the items of confectionery contained in the package minus the weight of immediate wrapper.

Batch Coding

A distinctive batch number or lot number or code number, either in numerical or alphabets or in combination, representing the batch number or lot number or code number being preceded by the words Batch No. or Batch or Lot No. or Lot or any distinguishing prefix. In case of canned food, the batch number may be given at the bottom, or on the lid of the container, but the words Batch No., given at the bottom or on the lid, shall appear on the body of the container.

Manufacturing Date and Best Before

The month and year in which the commodity is manufactured or pre-packed. Date of minimum durability This date consists of the day, month and year, except in the case of foodstuffs that will not keep for more than three months (the day and month are sufficient), foodstuffs which will not keep for more than 18 months (the month and year are sufficient), and foodstuffs which will keep for more than 18 months (year is sufficient).

Provided further that in case of package or bottle containing sterilised or Ultra High Temperature treated milk, soya milk, flavoured milk, any package containing bread, dhokla, bhelpuri, pizza, doughnuts, khoa, paneer, or any uncanned package of fruits, vegetable, meat, fish or any other like commodity, the declaration be made as follows:

BEST BEFORE.................DATE/MONTH/YEAR

OR

BEST BEFORE .... DAYS FROM PACKAGING

OR

BEST BEFORE ....DAYS FROM MANUFACTURE

OR

BEST BEFORE UPTO.... DATE/MONTH/YEAR

OR

BEST BEFORE WITHIN ....DAYS FROM THE DATE OF PACKAGING/MANUFACTURE

- Blank be filled up.
- Month and Year may be used in numerals.
- Year may be given in two digits.

The date of durability is not required for the following products:

- Untreated fresh fruits and vegetables.
- Wines and beverages containing 10 % or more by volume of alcohol.
- Non-alcoholic soft drinks.
- Fruit juices and alcoholic beverages in individual containers of more than five litres, intended for supply to mass caterers.
- Bakers' or pastry cooks' wares which are normally consumed within 24 hours of their manufacture.
- Vinegar.
- Cooking salt.
- Solid sugar.
- Confectionery products consisting almost solely of flavoured and/or coloured sugars.

- Chewing gums and similar chewing products.
- Individual portions of ice-cream.

In the case of foodstuffs which are highly perishable, the date of minimum durability shall be replaced by the 'use by' date.

Principal Display Panel, its Area, Size and Letter

- Principal Display Panel means that part of a label which is intended or is likely to be displayed, presented or shown or examined by the customer under normal and customary conditions of display, sale or purchase of the commodity of food contained in the package.
- The area of the principal display panel shall not be less than:
    - In the case of a rectangular container, forty percent of the product of height and width of the panel of such container having the largest area.
    - In case of cylindrical or nearly cylindrical, round or nearly round, oval or nearly oval container, twenty percent of the product of the height and average circumference of such container.
    - In the case of a container of any other shape, twenty percent of the total surface area of the container except where there is label, securely affixed to the container such label shall have a surface area of not less than ten percent of the total surface area of the container.
- In computing the area of the principal display panel, the tops, bottoms, flanges at top and bottoms of cans, and shoulders and necks of bottles or jars shall be excluded.
- In the case of package having a capacity of five cubic centimeters or less, the principal display panel may be card or tape affixed firmly to the package or container and bearing the required information under these rules.
- The height of any numeral in the declaration required under rules, on the principal display panel shall not be less than as shown in table.

Table: Height and Size of Letters to be printed for net quantity declared in terms of weight or volume on label.

Sr. No.	Net quantity in length, area or number, Area of Principal display panel	Net quantity in weight/volume	Minimum height of numeral in mm	
			Normal case	When blown, formed, moulded or perforated on container
1	Up to 100 cm^2	Up to 50g/ml	1	2
2	Above 100 cm^2, Up to 500 cm^2	Above 50g/ml up to 200g/ml	2	4

3	Above 500 cm^2, Up to 2500 cm^2	Above 200g/ml up to 1kg/litre	4	6
4	Above 2500 cm^2	Above 1kg/litre	6	8

Miscellaneous Information

- Languages of the Particulars or Declaration on the Label: The particulars of declaration required under these rules to be specified on the label shall be in English. Provided that nothing herein contained shall prevent the use of any other language in addition to the language required under this rule.
- Declaration to be surrounded by Line: There shall be a surrounding line enclosing the declaration and where the words "unsuitable for babies" are required to be used there shall be another such line enclosing these words. The distance between any part of the words "unsuitable for babies" and the surrounding line enclosing these words shall not be less than 1.5 mm.
- Labels not to Contain False or Misleading Statements: A label shall not contain any statement, claim, design, device, fancy name or abbreviation which is false or misleading in any particular concerning the food contained in the package, or concerning the quantity or the nutritive value or in relation to the place of origin of the said food.

Barcode

A barcode may be defined as a series of bars and spaces arranged according to the encodation rules of a particular specification in order to represent data. Its purpose is to represent information in a form that is machine-readable. Barcodes are read by scanning devices that are programmed to analyse the structure of the bars and spaces and transmit the encoded data in electronic format. These data can then be stored on a file or transmitted to a computer for processing. Techniques other than barcodes achieve the same objective: capturing automatically data encoded using a particular technology. These include optical character recognition, magnetic stripe, and radio-frequency identification. The concept of encoding and reading data automatically is called automatic data capture (ADC).

Benefits of Barcodes

The main benefits of barcodes are speed and accuracy. Capturing data automatically by reading a barcode can be done in a fraction of a second, much faster than manual key entry. It is commonly agreed that an operator doing key entry makes one error for every 300 characters typed. Reading barcodes makes data capture almost error-free. The error rate depends on the type of barcode and equipment being used, but usually it is lower than one error per 1,000,000 readings.

Barcode Symbologies

A barcode symbology is a set of rules describing the way bar and spaces have to be organized to encode data characters. Since the invention of the barcode concept in the United States in the late 1950s, hundreds of barcode symbologies have been developed, but only a few of them are actually being used on a large scale.

Typically, a symbology is qualified as being discrete or continuous. In a discrete symbology, the spaces between symbol characters do not contain information because each character begins and ends with a bar. In a continuous symbology, there is no intercharacter gap; that is, the final element of one symbol character abuts the first element of the next symbol character, and all the elements carry data contiguously.

Code 39

Code 39 was launched in 1975. It is widely used for industrial applications. Code 39 is a discrete, variable length symbology encoding the 36 numeric and uppercase alpha characters (A–Z, 0–9) and seven special characters: space, dollar sign ($), percent (%), plus (+), minus (–), dot (.), and slash (/). A symbol character is composed of nine elements, five bars and four spaces. An element is either wide or narrow. There are three wide elements and six narrow elements in a symbol character. A Code 39 symbol begins with a start character and ends with a stop character. It can be read from the right to the left and from the left to the right.

ITF

Interleaved two of five (abbreviated ITF) has been found to be well adapted to the materials and printing conditions frequently used on fiber-board cases. It is a continuous symbology encoding only numeric digits. A pair of digits is represented by five bars and five spaces. One of the pair is represented by the dark bars and the other by the light bars, and the dark and light bars are interleaved. Because the digits are represented in pairs, the symbol can only encode even number of digits. In addition to the digit characters, there are two auxiliary characters used as guard bars at the beginning and at the end of the digit representation. The symbol is designed to be read bi-directionally by fixed or portable scanners.

Code 128

Code 128 was introduced in 1981 in response to the need for a compact alphanumeric code symbol that could be used to encode complex data. The fundamental requirement called for a symbology capable of being printed by existing data-processing printers. Code 128 uniquely addresses this need with the most compact, complete, alphanumeric linear symbology available. In addition, Code 128 has been designed with geometric features to improve scanner reading performance and to be self-checking.

EAN/UPC

EAN/UPC was developed in the late 1960s when researches were conducted in the United States to improve the efficiency of checkout operations in retail stores. EAN/UPC is a continuous symbology encoding fixed-length numeric digits. Several variants exist, known as EAN-13, UPC-A, EAN-8, and UPC-E. In addition, the symbology enables to encode two small symbols encoding two and five digits. These are called add-ons because the information they contain supplement the main symbols. A symbol character is composed of seven modules, two bars, and two spaces. A bar or a space is composed of one to four modules. An EAN/UPC symbol begins and ends with a guard pattern. In the EAN-13, UPC-A, and EAN-8 version, a center pattern separates the symbol into segments that can be read separately by a decoding equipment, thus making the symbol omnidirectional readable. The EAN/UPC symbology is widely used to encode the identification number of consumer products.

PDF417

PDF417 is a two-dimensional, stacked bar-code symbology. In PDF417, the basic data unit or minimum segment containing interpretable data is called a code-word. Every code-word in the symbol is the exact same physical length, and each code-word can be divided into 17 equal modules. Within every code-word, there are four bars and four spaces. The minimum number of modules of any bar or space is one; the maximum is six. The PDF417 symbology defines 929 distinct code-words and supports 12 modes. Each mode specifies the meaning of the code-words. The standard modes are Extended Alpha Numeric Compaction Mode, Binary/ASCII Plus Mode, and Numeric Mode. The number of data characters that can be encoded in a PDF417 symbol depends on the mode being used. In the extended alphanumeric compaction mode, the maximum number of ASCII characters per symbol is 1850. In numeric mode, a symbol can encode a maximum of 2725 digits.

Data Matrix

Data Matrix is a two-dimensional matrix symbology that is made up of nominally square modules arranged within a perimeter finder pattern. Each Data Matrix symbol consists of data regions that contain nominally square modules set out in a regular array. The data region, or set of data regions and alignment patterns, is surrounded by a finder pattern, and this is in turn surrounded on all four sides by a quiet zone border. The number of data characters per symbol is up to 2335 alphanumeric characters or 3116 numeric only characters. Data Matrix Symbols are read by two-dimensional imaging scanners or vision systems. Most other scanners that are not two-dimensional imagers cannot read Data Matrix. Data Matrix Symbols are designed for use with applications that involve imaging scanners throughout the supply chain.

Applications

The bar-coding technology has gained wide acceptance in numerous applications.

Today, virtually all packages, from the smallest units intended for sale to a consumer to the biggest transport units, bear one or several barcodes carrying their identification number and other data relevant to the parties shipping, carrying, or receiving goods. Scanning at retail point of sale is a major application using bar-code technology. Millions of stores around the world have implemented scanning systems relying on the GS1 identification number and the associated EAN/UPC bar-code symbol.

A rapidly growing field of applications using bar-coding technologies lies within the supply chains. Goods ready for shipment by a supplier are packed, and each package is numbered and bar-coded with a unique number. Before the physical delivery of the merchandise, the supplier sends an electronic message to the delivery point, advising about the arrival of the goods. This electronic message contains the unique identification number of each package and the description of its contents.

References

- Cartoning-machine: saintytec.com, Retrieved 22, March 2020
- CANNING-MACHINERIES-AND-EQUIPMENTS-46104: ecoursesonline.iasri.res.in, Retrieved 10, May 2020
- How-does-stretch-wrapping-work: lantech.com, Retrieved 29, January 2020
- White-paper-orion-specifying-the-ideal-machine-20929: cdn2.hubspot.net, Retrieved 25, July 2020
- Explore-shrink-tunnels: industrialpackaging.com, Retrieved 18, February 2020

3

Food Packaging Technologies

There are several types of food packaging technologies such as active and intelligent packaging, aseptic packaging systems, antimicrobial food packaging and modified atmosphere packaging. This chapter discusses in detail the theories and concepts related to food packaging technologies.

Active and Intelligent Packaging

Active packaging can be defined as "Packaging in which supplementary constituents have been deliberately added in or on either the packaging material of the package headspace to improve the performance of the package system". In simple terms, Active packaging is an extension of the protection purpose of a package and is commonly used to protect against oxygen and moisture. It permits packages to interact with food and environment. Thus, play a dynamic role in food preservation. Developments in active packaging have led to advances in many areas, like delayed oxidation and controlled respiration rate, microbial growth and moisture migration. Other active packaging technologies include carbon dioxide absorbers/emitters, odour absorbers, ethylene removers, and aroma emitters.

On the other hand, intelligent packing can be defined as "Packaging that has an external or internal indicator to provide information about characteristics of the history of the package and/or the quality of the food packed". Intelligent packaging is an extension of the communication purpose of conventional packaging and provides information to the consumer based on its ability to sense, detect, or record external or internal changes in the product's environment. Example: Time-temperature indicators, ripeness indicators, biosensors, and radio frequency identification. Active and intelligent packaging are:

- Active packaging changes the condition of the packed food to extend shelf-life or to improve safety or sensory properties, while maintaining the quality of the packaged food.
- Intelligent packaging systems monitor the condition of packaged foods to give information about the quality of the packaged food during transport and storage.

Active Food Packaging Techniques

Active packaging technologies have started to receive a great deal of attention since the last decade. The market for active packaging films was about $50 million worldwide in 2004, and this market is growing rapidly. Active packaging techniques for preservation and improving quality and safety of foods can be divided into three categories; absorbers or scavengers, releasing systems and other systems. Absorbing (scavenging) systems remove undesired compounds such as oxygen, carbon dioxide, ethylene, excessive water, and other specific compounds. Releasing systems actively add or release compounds to the packaged food or into the head-space of the package such as carbon dioxide, antioxidants and preservatives. Other systems may have miscellaneous tasks, such as self-heating, self-cooling and preservation.

Depending on the physical form of active packaging systems, absorbers and releasers can be a sachet, label or film type. Sachets are placed freely in the head-space of the package. Labels are attached into the lid of the package. Direct contact with food should be avoided because it impairs the function of the system and, on the other hand, may cause migration problems.

Oxygen Scavengers

High levels of oxygen present in food packages may facilitate microbial growth, off-flavours and off-odours development, colour change, and nutritional losses, thus causing significant reductions in the shelf-life of foods. Thus, the control of oxygen levels in packages is important to limit the rate of such deteriorative and spoilage reactions in foods. Oxygen absorbing systems provide an alternative to vacuum and gas flushing and improve product quality and shelf-life. Also, they are economically viable in reducing packaging costs and increasing profitability.

Typical oxygen absorbing systems are based on the oxidation of iron powder by chemical means or scavenging of oxygen by the use of enzymes. In the former, iron kept in a sachet is oxidized to iron oxide. The sachet material is highly permeable to oxygen and in some cases, to water vapour, for the sachet to be effective. This well-known oxygen scavenging system was first developed and introduced to the food packaging market by a company. The type and amount of absorbent that needs to be used in a sachet is determined by the initial oxygen level in the package, the amount of dissolved oxygen present in food, permeability of the packaging material, nature (size, shape, weight, etc.) of food and the water activity of the food. These iron-based oxygen absorbing systems have the ability to scavenge oxygen in many foods, including high, intermediate, or low moisture foods, and foods containing lipids. They can also work at refrigerated and frozen storage conditions, and can be used as effective oxygen scavengers with microwaveable food products.

In enzymatic oxygen scavenging systems, an enzyme reacts with a substrate to scavenge oxygen. These systems are more expensive than iron-based systems, due to the cost of enzymes used for the oxygen scavenging purpose. Enzymatic oxidation systems are also usually very sensitive to temperature, pH, water activity and solvent/substrate present in the sachet, thus limiting the widespread use of these enzyme-based systems.

Oxygen scavenging sachets are not appropriate for liquid foods, because the direct contact of the liquid with the sachet usually causes the leakage of sachet contents. Also, sachets may cause accidental consumption with the food or may be ingested by children. Oxygen scavenging sachets sold in the US are required to be labelled "Do not eat," for safety reasons and regulatory purposes mandated by the Food and Drug Administration. Although sachets can be concealed using secondary packages, but this practice increases packaging costs.

The incorporation of scavengers in packaging films is a good way of resolving sachet-related problems. Scavengers may be inserted into a solid, dispersed in the plastic, or inserted into various layers of the package, including adhesive, lacquer, or enamel layers. Multi-layer oxygen scavengers more effectively absorb oxygen than single layer scavenging systems.

In the structure of a typical multi-layer oxygen scavenging system, an oxygen absorbing substance imbedded in a layer very permeable to oxygen is used to absorb oxygen present in the package head space. Oxygen entry from the outside environment to the oxygen absorbing layer is limited by a barrier layer, which is highly impermeable to oxygen. An inner layer, or a control layer, next to the oxygen absorbing layer is used to minimize any migration of the oxygen absorbing substance into the food.

Scavengers in the form of films allow the absorption of oxygen from all surfaces of the food that are in contact with the film. The rate of oxygen absorption by films changes, depending on the area and thickness of the film. A Corporation has developed a polymer based film that can overcome the negative effects of iron-based oxygen scavenging films. This film uses innovative technology that is invisible to consumers because the scavenging component is co-extruded as a layer of the package. Since oxygen absorbing material is virtually invisible in the films, these films do not alter the look of the package. They also offer shoppers a clear view of the product packed. It is equally effective with wet and dry products, because its scavenging action is initiated "on demand" on the processor's packaging line by an ultraviolet-light triggering process.

Carbon Dioxide Scavengers and Emitters

High levels of carbon dioxide usually play a beneficial role in retarding microbial growth on meat and poultry surfaces and in delaying the respiration rate of fruits and vegetables. For instances where the package has a high permeability to carbon dioxide, a carbon dioxide emitting system may be necessary to reduce the rate of respiration and suppress microbial growth. The use of a dual function system consisting of an oxygen scavenger and a carbon dioxide emitter is the usual practice for increasing the shelf life of highly perishable foods.

On the contrary, dissolved carbon dioxide formed after the roasting of coffee may cause the package to burst, if the roasted coffee is packed in a can or aluminium foil pouch. This released carbon dioxide from freshly roasted coffee can be scavenged through the use of a carbon dioxide scavenger. A carbon dioxide absorbing sachet is developed, which is composed of a porous envelope containing calcium oxide and a hydrating agent, such as silica gel, on which water is adsorbed. In this system, water reacts with calcium oxide and produces calcium hydroxide, which then reacts with carbon dioxide to form calcium carbonate.

Moisture Scavengers

If the package has a low permeability to water vapour, water accumulation inside the package is more pronounced. The excess water development inside a food package usually occurs due to the respiration of fresh produce, temperature fluctuations in high equilibrium relative humidity food packages or the drip of tissue fluid from cut meats, poultry, and produce.

An effective way of controlling excess water accumulation in a food package that has a high barrier to water vapour is to use a moisture or water vapour scavenger, like silica gel, molecular sieves, natural clays, calcium oxide, calcium chloride and modified starch, or other moisture absorbing substances. Silica gel is the most widely used desiccant because it is nontoxic and non-corrosive.

A blanket or a pad of desiccant is usually wrapped around the food to be preserved (meat, poultry or fish) to absorb water. Another approach to absorb moisture from these foods is to use a superabsorbent polymeric laminate film that has a moisture absorbent layer that is formed from a polyester graft copolymer and a resin, consisting of a polyurethane resin, an acrylic resin and vinyl resin. Moisture absorbing systems in sachet are usually used to maintain low levels of moisture in dried food packages, like chips, nuts, spices, biscuits, crackers, milk powders and instant coffee. The sachets like Desi Pak, Desi View, Sorb-It and 2-in-1 from United Desiccants, USA, MiniPax, StripPax, Natrasorb, and the moisture absorbing label Desimax from Multisorb Technologies, USA are the most common moisture absorbing systems used to absorb and/ or control moisture for water sensitive packaged foods in USA.

Ethylene Absorbers

Ethylene is a hormone that accelerates ripening and senescence by increasing the respiration rate of climacteric fruits and vegetables, thereby decreasing shelf-life. Hence, the removal of ethylene gas from the package headspace slows senescence and improves shelf-life. The most well-known, inexpensive, and extensively used ethylene absorbing system consists of potassium permanganate inserted in silica. The silica absorbs ethylene, and potassium permanganate oxidizes it to ethylene glycol. Silica is kept in a sachet permeable to ethylene, or it can be incorporated into a packaging film. Potassium permanganate is not integrated into food contact surfaces of packaging films due to its toxicity. The substrate surface area and the amount of potassium permanganate affect the performance of these systems.

Another system is based on infusing zeolite with potassium permanganate, and then coating the infused zeolite with a quaternary ammonium cation. This system is not only capable of absorbing ethylene from the medium, but also other organic compounds, such as benzene, toluene, and xylene. Ethylene scavengers have been proven to be effective in the storage of packaged fruits, including kiwifruit, bananas, avocados, and persimmons. Ethylene scavengers are commercially available as Evert-Fresh (Evert-Fresh Co, USA), Ethylene Control (Ethylene Control Incorporated, USA) and Peakfresh (Peak Fresh Products, Australia).

Flavour Absorbing/Releasing Systems

Food packaging materials like some plastics may interact with food flavours, resulting in loss of flavours, known as flavour scalping. Flavours are usually lost or degraded after processing foods at high temperatures or after packaging. Therefore, there is a need to replace these lost flavour constituents when scalping occur. In addition, consumers always like to smell good flavours when they first open a food package. For example, most of the dry instant coffee manufacturers often fill the headspace with distilled volatiles to have fresh coffee fragrance when the package is first opened.

In contrast to flavour releasing systems, flavour absorbers scavenge undesirable flavours, aromas, and odours present in the package headspace. Some orange varieties, such as Navel may develop a bitter flavour when limonin concentration exceeds 12 mg/kg after pasteurization. Internally cellulose acetatebutyrate coated plastic bottles, which act as limonin absorbers, have been shown to considerably reduce bitterness in orange juice thereby increasing shelf-life.

The formation of off-flavours and off-odours in food products are due to (1) the oxidation of fats and oils, leading to the formation of aldehydes; (2) the breakdown of proteins of fish muscle into amines. Although aldehydes and amines can be removed from package headspaces by flavour scavengers, these systems may mask or absorb off-flavours and off-odours that are indicative of spoilage.

Intelligent Packaging Techniques

The definition of intelligent packaging in the Actipak project includes indicators to be used for quality control of packed food. They can be so-called external indicators, i.e., indicators which are attached outside the package like time-temperature indicators, and so-called internal indicators which are placed inside the package, either to the head-space of the package or attached into the lid like oxygen indicators for indication of oxygen or package leak, carbon dioxide indicators, microbial growth indicators and pathogen indicators.

Table: Examples of internal and external indicators.

Indicator	Principle/reagents	Gives information about	Application
Time-temperature indicators (external)	Mechanical, Chemical, Enzymatic	Storage conditions	Foods stored under chilled and frozen conditions.
Oxygen indicators (internal)	Redox dyes, pH dyes Enzymes	Storage conditions Package leak.	Foods stored in packages with reduced oxygen concentration.
Carbon dioxide indicator (internal)	Chemical	Storage conditions, Package leak	Modified or controlled atmosphere food packaging.
Microbial growth indicators (internal /external) i.e. freshness indicators	pH dyes, All dyes reacting, with certain metabolites (volatiles or non-volatiles).	Microbial quality of food (i.e. spoilage).	Perishable foods such as meat, fish and poultry.
Pathogen indicators (internal)	Various chemical and immunochemical methods reacting with toxins.	Specific pathogenic bacteria such as Escherichia coli O157.	Perishable foods such as meat, fish and poultry.

Time-temperature Indicators

Temperature abuse is common during storage, transportation, and distribution; these indicators are designed to monitor temperature abuses in a food product's shelf life. Temperature abuse does not only cause quality and nutritional losses, but also may lead to food poisoning and food losses. Many time-temperature indicating systems are based on a colour change or colour development, which is correlated with food quality loss. Nevertheless, time-temperature indicators that have a response in the form of visible mechanical deformation are also available. Time-temperature indicators must satisfy some requirements to be effective as monitoring devices such as:

- They must be easily activated and sensitive.
- They must provide a high degree of accuracy and precision.
- They should have tamper evident characteristics.
- Response should be irreversible, reproducible, and correlated with food quality changes.
- Physical and chemical characteristics of time-temperature indicators should be determined.
- Response should be easily readable and, should not be confusing.

Time-temperature indicators for frozen, refrigerated, modified atmosphere packaged and thermally processed foods are commercially available. Although time-temperature indicators are still in their infancy, continued research and consumer appreciation are expected to play positive roles in the development of more reliable and sophisticated systems.

Freshness Indicators

A freshness indicator indicates directly the quality of the product. The indication of microbiological quality is, for example, based on a reaction between the indicator and the metabolites produced during growth of microorganisms in the product. The freshness indicators are mainly based on compounds like glucose, organic acids and their salts, ethanol, volatile nitrogen compounds, carbon dioxide and sulphuric compounds. For example, Concentration of D-lactate increases with storage of meat. Thus it can be taken as freshness indicator.

An indicator that would show specifically the spoilage or the lack of freshness of the product, in addition to temperature abuse or package leaks, would be ideal for the quality control of packed products. The number of concepts of package indicators for contamination or freshness detection of food is still very low, however new concepts of freshness indicators are patented and new commercially available products are likely to become available in the near future.

Other Active and Intelligent Packaging Systems

Other active packaging applications that are expected to find increased consideration in the future include antioxidant releasing films, colour containing films, light absorbing/regulating systems, anti-fogging and anti-sticking films, susceptors for microwave heating, gas permeable/breathable films, and insect repellent packages. For example, antioxidant incorporated films can be used to prevent the oxidation of fats and oils that lead to rancidity. They can also be used to prevent the formation of off-odours and off-flavours in foods.

Aseptic Packaging Systems

Aseptic packaging is a relatively new packaging concept aimed at developing new product group, namely pre-sterilized and aseptically packaged foods. During aseptic packaging process, a pre-sterilized product is filled under sterile conditions into sterile packages in a sterile environment and then germproof sealed. Although both the aseptic can filling and aseptic carton filling systems became commercial during the late fifties and early sixties, only the latter found application in aseptic packaging of liquid foods, more particularly UHT milk and fruit juices. During the past few years, environmental considerations have led to the use of recyclable glass bottles instead of cartons in countries like Germany.

Requirements of Aseptic Packaging Systems

The major requirement of an aseptic packaging unit is to prevent recontamination of the sterilized product. The principal considerations in this regard include sterilization of the filling machine and packaging material by suitable physical and/or chemical means and maintaining aseptic barriers during filling and sealing. Besides the equipment and packaging, gas used to pressurizing the filling space is one of the sources of recontamination of processed foods. Thus mechanical failures such as inadequate heating of the gas, leaks in valves and pinholes in filters may cause recontamination and must, therefore, be checked. Types of packaging materials for aseptic packaging systems and their properties:

- Glass: Glass bottles have been used for packaging of many liquid foods but it has not been a commercial success with aseptic filling plants. Glass as such offers protection against oxygen and light if it has been provided with anti-actinic compounds.
- Cans: Aseptic canning is expensive, particularly for a low cost product like milk. Cans are not preferred for packaging of UHT processed products as the processors and marketers of the product generally want to emphasise the newness of the process whereas cans are identified with conventionally retorted products. The cans may be of tinplate or drawn aluminium: the solder in tinplate cans may have to be of higher melting point than normal to withstand the can sterilization temperatures.
- Paper board cartons: Such packaging materials are commonly used in aseptic filling systems for milk, cream, fruit juices, soups, etc. The filling systems could be either of the following two types: those in which the carton is formed within the filler from a continuous reel of materials; and those in which the cartons are supplied as performed blanks, folded flat, which are assembled into cartons in the filler. The packaging material is Food Packaging mainly composed of printed-paper coated with aluminium foil and several plastic layers (polyethylene-paper board-polyethylene-aluminium foil-polyethylene). The inner material side of the finished package is coated with a special layer facilitating the sealing process. Each layer has a specific function:
    - The outer polyethylene layer protects the ink and enables the sealing process of the package flaps.
    - The paperboard serves as a carrier of the décor and gives the package required mechanical strength.
    - The laminated polyethylene binds the aluminium to the paper.
    - The aluminium foil acts as a gas and light barrier.
    - The inner polyethylene layer provides liquid barrier and enables sealing of the package.

Sterilization of the Packaging and the Filler Environment

Chemical sterilization processes for the packaging film include treatment with ethylene oxide, sodium hypochlorite, peracetic acid and hydrogen peroxide (H_2O_2). Ethylene oxide is not only slow in action but it's desorption requires very long time. Hence it can be used for pre-treatment of packaging, but not for final sterilization on the packaging unit. Sodium hypochlorite and peracetic acid are very effective sterilants, but removal of their residues from the packaging necessitates a sterile-water rinse. Alcohols such as glycols require high temperature (e.g. 100 degree C) application for the desired sporicidal effect. Although H_2O_2 also shows poor effectiveness at ambient temperatures, its high sporicidal effect at 80 degree C makes it useful for packaging sterilization. It is first applied on the material and then evaporated by heating through hot air or infrared radiation. The limitations of the use of H_2O_2 are: (1) the surfactant/surfactants or wetting agents used for uniform deposition on the packaging film, cannot be evaporated by heat and thus may find their way into milk, (2) the vapours of H_2O_2 must be exhausted to avoid injury to the workers, and (3) the efficacy of its removal by evaporation must be monitored through routine testing of milk.

While steam or hot water is effective in sterilization of the milk carrying tubes, hot air (300 degree C) with or without filtration, is commonly used for sterilization of the air injected in the filling space. Air at 330-350 degree C (for 30 mins) may also be used for milk tube sterilization. Sterilized at 180-200 degree C is used to evaporate H_2O_2 and when cooled to 50 degree C can be employed for pressurizing the filling chamber.

Effective use of UV radiation imposes certain stringent requirements such as perpendicular incidence of rays, dry atmosphere, smooth surface, low concentration of microorganisms, and absence of visible light to avoid reactivation of microorganisms and shields to protect the operator. Therefore, UV radiation is suitable only for a complementary treatment of already sterilized packaging. Filtration by means of depth filters (mats of compressed glass- or asbestos-fibre, or of sintered metal or ceramic) is effective in freeing air from bacteria. The filters themselves may be sterilized by fumigation, hot air or steam. Aseptic barriers in the form of steam or circulated liquid sterilant become necessary with valves and fittings coming in contact with sterile milk. Detection of leaks by using a dye test is imperative to check recontamination of the packaged sterile milk.

Aseptic Packaging Systems

Filling of commercially sterile milk and fruit juices in sterilized packages/containers in a sterile environment, and hermetically sealing the same to prevent recontamination of the milk can be achieved in two major ways: (a) using presterilized preformed containers such as bottles and cans, and (b) sterilizing the packaging material, forming it into suitable containers, filling the sterile product and sealing the package on the so called form-fill-and-seal (FFS) machines. The latter employs a multiply laminate of polyethylene, polystyrene and/or polypropylene films, paper and aluminium foil.

The Dole aseptic canning system has been used for UHT milk in the USA, but only to a limited extent. The most widely used FFS Tetrapak systems using tetrahedron cartons, and tetrabriks or hexahedron cartons are characterized by continuous formation of the package below the milk level from a paper/PE/Al laminate strip which has been continuously sterilized by H_2O_2 boiled off by radiant heat in the region immediately above the milk surface thus giving a sterile atmosphere in the packaging zone. Recently, tetra pak has introduced the so called 'pillow pak' to cut down the packaging cost of UHT milk.

Coupling of Aseptic Packaging with the UHT Plant

In small processing units, a single flow-sterilizing plant can be connected with a single packaging plant of the matching capacity. But this system is inflexible because both the sterilizer and filler must operate simultaneously. If one stops for any reason, the other must be shut down, or in case of the filler stopping, the sterilized products must be recirculated for reprocessing. This problem can be solved by providing an aseptic tank at the interface between the two plants.

In large units, it is a general practice to feed two or more fillers from a single sterilizing plant whose capacity is equal to the total of filling capacity. So, if one of the fillers has to be shut down only a small portion of the UHT processed products will be required to be recycled. Use of a variable speed homogenizer can altogether eliminate the need of recirculation of sterilized product in such a situation. Even with a multiple filler system, aseptic tank is very useful for smooth operation of the whole system without jeopardizing the product quality. Aseptic balance tank, however, adds to the investment cost as well as cleaning and sterilization requirements. It also requires a supply of sterilized air for partial positive pressure during its use.

Retort Packaging for Long Shelf-life Foods

The concept of retort pouch which took shape in the 1940's, was developed again in response to the military need, as the rigid cans conventionally used by the combat forces posed problems such as difficulty in opening, injurious, and the potential to be used in makeshift explosives by the enemy. The retortable flexible packages are characterized by their structural components of heat resistant plastic layers with or without aluminium foil and their ability to be thermo processed to result in shelf-stable food product. Its cost is also less as compared to cans. An early pioneer in proving the production reliability of the retort pouch was the US Army Natick Research and Development Center. In many developed countries, most ready-to-eat foods are packaged in retort pouches. Japan has been a pioneering country in this respect. In most recent times, focus has shifted to retortable semi-rigid trays or tubs because of their added convenience in use.

Retort Pouch Material

The materials used in making retort pouches should possess toughness and puncture resistance properties normally required of flexible packaging, good barrier properties for long shelf-life, and heat sealability over a wide temperature range along with the ability to with stand processing temperatures of the order of 110-140 degree C. To have all of these properties, laminated structures or co-extruded films are used.

The outer film of the composite structure is needed for strength and resistance. It should be resistant to heating temperatures, printable and be able to withstand temperatures without bursting, shrinking and delamination. The most common material used in polyethylene terephthalate (PET). It has the added advantage of being reverse printed so that ink is embossed between the outer layer and the next inner layer. In order to achieve a shelf life of one or more years, aluminium foil layer as one of the inner layers for barrier properties is essential. The thickness range of aluminium foil varies from 9 to 25μm, though a thickness of 9-10μm is most common. In Japan, retort pouches without aluminium foil layer are also very common as some products with a low shelf life of 3-6 months are also acceptable. Nylon is another material used as a barrier film in place of aluminium foil because of its low gas transmission rate and toughness. However, being transparent, nylon based laminates cannot provide protection from light unless covered with a carton or wrap. The current material most commonly used as innermost sealant layer is cast polypropylene, though high density polyethylene modified with isobutylene rubber has also been used.

Sterilization Methods for Packaging Materials and Equipment

Various methods for the sterilization of packaging materials are currently used in aseptic packaging systems. The sterilization process employed should be established in terms of numbers of log cycle reductions of the most resistant organisms. Packaging material is usually sterilized either inside the packaging machine or externally and introduced aseptically into the aseptic zone of the packaging machine. Many aseptic packaging systems utilize plastics, paperboard or foil containers to pack the sterilized product and sterilize by physical methods and chemical reagent.

Thermal Processes

Microorganism inactivation has traditionally been accomplished by heating. Microorganisms, especially spores, show greater thermal resistance when exposed to dry heat than moist heat. Spores of Bacillus stearothermophilus are extremely resistant to heat. Spores of Bacillus subtilis ATCC 9372 are used as indicators in dry heat and these spores are used commercially in sterility testing of aseptic fillers. D-values (the time required to reduce the number of microorganisms by one log cycle) and Z-values (the temperature rise necessary to reduce the D-value by a factor of 10) are significantly higher for dry heat sterilization than for steam sterilization. Dry heat produces microbial death as a result of dehydration followed by protein oxidation. Death by moist heat is caused by denaturation and coagulation of essential cell proteins. Reaction velocity for their destruction depends on how rapidly the heat from the thermal carrier can be transferred to the cell. When heat is used, the nature of the surface must be considered. Plastics or carton packaging with their low conductivity are more difficult to thermally sterilize than are metal containers. In addition, plastic materials generally have a low thermal stability and can be permanently deformed by the time=temperature sequences necessary to achieve sterilization. Thermal processes do not deposit any hazardous or undesirable residues on the surface being treated and do not present environmental hazards.

Saturated Steam

Saturated steam is a form of moist heat. Sterilization of metal cans and lids by saturated steam under pressure was used as early as 1920 in the USA and is used today for sterilizing thermostable plastic cups. To attain adequate destruction in the short time available in high speed packaging units, the surface temperature of the material must reach 135 °C. The technological problem is to obtain a sufficiently high surface temperature to achieve the required sterility in a time consistent with high production rates and to avoid softening and deformation of the material. Moulded polystyrene cups and foil lids are subjected to saturated steam at 165 °C and 6 bars immediately after deep drawing. At the same time the external cup surface is cooled to limit the effect of the short time heat application on the material. It should be noted that Bacillus subtilis is not generally recommended for validation of steam sterilization. Polypropylene cups that have higher thermal tolerance can be treated on surfaces at lower temperatures and longer times, 140–147 °C for 4–6 s, in a pressure chamber. Lid foil can be treated continuously by passage through a pressure lock. Moist heat could cause blistering or delamination of paper-based packaging materials and impair the heat-sealing characteristics of plastics. Atmospheric steam can only be used on non-paper-based pre-formed containers. Saturated steam is a preferred treatment method for sterilizing metal food contact surfaces downstream from the hold tube, including sterile hold tank, homogenizers, fillers and the aseptic packaging zone. Equipment sterilization can be attained by exposure of the surfaces to an appropriate time/temperature sequence (e.g. 30 min at a surface temperature of at least 121 °C) by superheated water, saturated steam, superheated steam or other appropriate treatments.

Superheated Steam and Hot Air

Dry heat delivered by superheated steam or hot air is not as effective as moist heat at the same temperature for sterilization, therefore higher temperatures are required. The Martin–Dole process continuously sterilizes tinplate cans by passage through 220–256 °C-superheated steams at normal pressure for 45 s. When paper-based packaging materials are used, hot air is preferred as a sterilant over superheated steam. Combicans of cardboard laminates (i.e. aluminium and paper) are sterilized in hot air at 145 °C for 3 min. This method has only been found suitable for products that have a pH of <4.6.

When sterilizing by hot air or superheated steam, there is no effect on packaging material quality. Furthermore, the temperature within the entire aseptic filling chamber may be maintained at a level lethal to microorganisms, thus assuring asepticity. The high temperature of the headspace gas at the time of sealing also reduces the level of oxygen in the sealed package and can contribute towards prolonging product shelf life. Hot air sterilization units are commercially available for juice and beverages. Dry heat sterilization is slow and not suitable for heat-sensitive materials like many plastic items.

Heating by Extrusion

Aseptic packaging system utilizing heat co-extrusion process for sterilization is a form–fill–seal packaging system that relies on the temperatures reached by thermoplastic resin, during the co-extrusion process used to produce multi-layer packaging material, to produce a sterile product contact surface. Temperatures of 180–230 °C can be reached during the extrusion of granulated plastics as a pre-treatment for a subsequent blow moulding operation. The high temperature produced by extrusion produces sterile surfaces. During production, the multi-layer package material is fed into the machine where it is delaminated under sterile conditions. This removes a layer of material and exposes the sterile product contact surface. The container material is then thermoformed into cups. The lid material that is also delaminated is then sealed onto the cup after filling. The sterility of the forming, filling and sealing areas is maintained by sterile air under positive pressures. Reports as to the efficacy of this heating method vary considerably. Therefore, aseptic filling into extruded containers appears only suitable for acidic products with pH less than 4.6. For products with pH greater than 4.6, extruded containers should be post-sterilized with hydrogen peroxide or with mixtures of peracetic acid.

Radiation

When neither heat nor chemicals can be used to sterilize a given material, radiation is considered as an option. Electromagnetic radiation characterized by a frequency, a wavelength, penetrating power and an energy range are infrared, ultraviolet, γ rays, etc. The dosage and type of radiation sterilization is a function of the type and amount of the microbial load that has to be removed from a given materials and equipment. As irradiation does not leave any residue on the treated surface nor affect the immediate environment, it appears at first sight to be a perfect solution for sterilization of packaging material. The products used for packaging in radiation sterilization recommended by the International Atomic Energy Agency (IAEA) are polyethylene, polyester, polypropylene, nylon, PVC, etc. Irradiation may indirectly affect chemical or biological properties, leading to interaction with the packaged product, changing its efficacy or usability.

UV Rays

UV irradiation has been successfully applied to the sterilization of air and of water which is free from suspended matter. With the development of more powerful sources of UV radiation, attempts have been made to use it for surface sterilization. The microbicidal effects of UV rays lies in the range of 200–315 nm, with an optimal effect being between 250 and 280 nm, the so-called UV-C range. UV light at a wavelength of 253.7 nm is an effective germicide against yeast, moulds, bacteria, viruses and algae. The lethal irradiation dose is defined as the product of irradiation intensity and time and expressed in terms of milli Watts second per square centimeter (mW s cm^{-2}).

High intensity sources of UV radiation are very effective in killing unprotected microorganisms suspended in air or on the surface of materials and products. Data published by Bachmann on dose levels required inactivating spores of moulds and bacteria indicated excellent potential of the process for sterilizing aseptic packaging materials. Researchers suggested the use of high-intensity UV for aseptic packaging material sterilization. Both vegetative cells and spores are sensitive to UV radiation. Microbial resistance to UV radiation increases in the order: vegetative bacterial cells ≪ yeasts < mould spores. Gram-positive bacteria require twice the dosage of Gram-negative bacteria, and 5–10 times for bacterial spores. Mould spores, particularly those that have dark colours (e.g. Aspergillus niger) are especially resistant and require dosage levels 20–100 times higher. Covered, shaded or shielded organisms will not be affected. Dust particles present on the surfaces reduce the effectiveness of UV irradiation for sterilization of aseptic packaging materials. The penetration ability of UV rays is very limited.

Table: Penetration of UV light.

Substrate	Distance that UV will penetrate (cm)
Air	300–500
Water	30
Window glass	<0.1
Transparent plastic foil	<0.01
Fruit juice	0.1
Fluid milk	0.01

A typical UV radiation intensity is between 0.5 and 1.5W cm^{-2} at a distance of 10 cm. Laboratory tests using an energy density of 30mW cm^{-2} on artificially contaminated smooth, even, dust-free surfaces and an irradiation time of 4–6 s showed a 4–5 log cycle reduction. Researcher obtained a patent for both the method (UV light plus organic acid or hot water) and apparatus for sterilizing packaging film. The reported advantage of this method is that organic acids and water are not toxic. UV irradiation and a citric acid solution resulted in approximately a six-log cycle inactivation of Bacillus subtilis spores within 5 s. Researcher described an aseptic packaging machine that used both an alcohol bath and high intensity UV radiation for sterilization of flexible plastic pouches. The film was washed in 95% ethanol for 20 s and then passed under a flow of 10 mW cm^{-2} of UV for 30 s. The rate of microbial reduction in the sterilization of smooth surface plastic cups and lids by UV irradiation at a constant radiator output depends upon a number of factors. UV irradiation under practical conditions only always count reduction of 2–3 log, and mould spores can be particularly resistant. This low count reduction is insufficient for aseptic packaging; the process is therefore merely suitable for a filling and packaging process with a relatively low contamination count.

Combination of Hydrogen Peroxide with UV

The ability of UV and hydrogen peroxide to inactivate bacteria has been well known for a long time. The process of simultaneously applying hydrogen peroxide solution and UV will provide adequate sporicidal activity to relatively low concentrations of hydrogen peroxide solution, minimizing the problem of removal of residues after sterilization. The Liquid-Pak System uses this effect with a carton forming, filling and closing system in a sterile chamber. Many researchers have considered sterilization of packaging materials using peroxide and UV irradiation. It has been found that the lethal action of peroxide solutions with and without the application of heat is increased by simultaneous UV irradiation, and that the overall lethal effect is greater than the sum of the effects of the peroxide and irradiation alone. They subsequently demonstrated the same type of synergy with a variety of Bacillus and Clostridium, but found that mild heat was also needed for effective killing of more resistant spores tested. Higher count reduction can be obtained by prior treatment of the surface with cold hydrogen peroxide and by ensuring that the surface is dust free.

The effect is optimum at a relatively low peroxide concentration, between 0.5 and 5% and at higher concentration the peroxide appears to have a protective effect, and the greater the UV intensity the higher is the optimum peroxide concentration. Four decimal reductions of Bacillus subtilis spores of a strain very resistant to H_2O_2 were easily obtained by irradiation with an UV dose of 1.8 W cm^{-2} in 2.5% peroxide followed by a short heat treatment. Subsequent heating to 80 °C increased the kill further. Maunder used a conventional UV lamp to sterilize pouches; also with the advent of high intensity UV-C lamp, very high lethal effects have been reported; up to 5 decimal reductions on at board and 7 decimal reduction for Bacillus spores on inoculated strips placed into food packaging cartons.

Decontamination of pre-formed cartons for aseptic packaging poses problems which are not encountered with flat surfaces. Therefore, UV irradiation of a pre-formed carton must be considerably more intense than for flat boards to ensure that all surfaces are effectively irradiated. Experiments with cartons artificially contaminated with Bacillus subtilis spores, sprayed with 1% H_2O_2 and then irradiated for 10 s with a high-intensity UV source above the carton, showed 5 decimal reductions with polyethylene lined material and 3.5 decimal reductions with a polyethylene=aluminium foil laminate. No heat was applied. This combination of hydrogen peroxide with UV has now been applied commercially to carton sterilization during aseptic filling.

Infrared Rays

Infrared rays in the waveband 0.8 – 15 × 10^{-6} are easily generated. On an absorbent surface they are converted into sensible heat, producing an increased temperature. The energy impact on the surface that is convertible into heat is dependent on the same geometric conditions as those outlined for UV rays. Thus, infrared irradiation is best applied to a smooth even surface with vertical radiation incline. This method has been used to treat the interior of aluminium lids with a plastic coating on the exterior surface. As the temperature rise can cause a softening of the plastic, the maximum temperature and exposure time are limited. Count reductions on the same order as those for UV irradiation has been found.

Ionizing Radiation

Ionizing radiation is an excellent sterilizing agent and penetrates deep into objects. Ionizing radiation of primary interest includes α-rays, X-rays, β-rays, electron beams and γ-rays. Bags used in bag-in-box aseptic packaging are currently sterilized with γ-rays from a ^{60}Co source. Radiation kills bacteria in the same logarithmic fashion as heat, and D-values for many organisms have been reported. A dose of 4.7 Mrad is considered a 12D level for Clostridium botulinum which is the most radiation-resistant spores of public health and spoilage significance. The effect of water activity on irradiation by ^{60}Co γ-rays was investigated by some researchers. Dosage levels to inactivate 90% of Bacillus subtilis spores increased from 0.23 to 0.42 Mrad as the water activity (a_w) decreased from 1.00 to 0.00. For a 99.9% reduction (3 log) of Bacillus stearothermophilus spores, the dosage increased from 1.7 Mrad at an a_w of 1.00 to 3.00 Mrad at 0.00.

Gamma irradiation has been used to sterilize plastic bags for bulk packaging acid foods using an aseptic bag-in-box system. The bag made of plastic laminates are treated in a specialized irradiation plant, and given a radiation dose of 25 kGy (2.5 Mrad) or more, which is sufficient to ensure sterility. However, the difficulty of maintaining sterility during the transfer operation to the filling spout of an aseptic filling machine, raises some doubts as to the practicality of gamma irradiation sterilized packaging materials for use on low-acid foods. The heavy shielding necessary to prevent radiation leakage from gamma ray source makes the system impractical to use for in-line sterilization. Gamma radiation has not yet been widely employed in the USA because of the cost and concerns about the effects of γ-radiation on food.

Electron beam sterilization of packaging material is a proven, state-of-the art technology, which outperforms any legislative and customer requirements. Studies have shown that depending on the level of microorganisms, a dose of 5–7 kGy was found to be effective against yeast, mould and spores and no viable microorganism was found after treatment. Unfavourable effects of electron beam treatment were not observed. High-energy electron beams are not as penetrating as γ-rays, and therefore do not require as much shielding as a γ-ray sources. The irradiation process takes just a few seconds and does not leave any type of residue. However, with pre-formed containers, the geometry of the delivery system makes it difficult to apply a uniform dose on the whole container. In-line processing, heavily demanded by industry, requires a careful process design. Technological and economic optimization of electron beam source and shielding are the most important tasks to be accomplished before in-line electron decontamination of food packaging may be applied on a large scale. A major impediment to adoption of this technology in the food industry is the high initial and operating cost of the units.

Care must be used when using ionizing radiation for sterilization of aseptic packaging materials. Although aluminium foil is not affected by irradiation, paper and plastic could be affected, with results ranging from discolouration to loss of pliability and mechanical strength. While radiation can be quite effective, the equipment is expensive, as is the protection for workers and the environment. Workers are not usually willing to work in such an environment. It will probably be more extensively employed in the future.

Light-pulse

The technology of utilizing short pulses of light is an attractive alternative for sterilizing packaging materials and processing equipment in aseptic packaging. The spectrum of light used for sterilization purposes includes wavelengths in the ultraviolet to wavelengths in the near-infrared region. The material to be sterilized is exposed to at least one pulse of light having an energy density in the range of about 0.01– 50 J cm^{-2} at the surface. The duration of pulses ranges from 1 μs to 0.1 s. The packaging materials are typically exposed to between 1 and 20 pulses of light intensity, short-duration light. A few flashes, applied at a rate of 1–10 per second, provide very high microbial kill levels, making the system ideal for continuous in-line sterilization. It is desirable for certain aseptic processes that packaging material be treated with pulses having a relatively high UV content to minimize the total influence (J cm^{-2}) necessary to achieve the desired reduction in microbial population. Comparison of the antimicrobial effects obtained using pulsed light with those obtained using non-pulsed or continuous wave conventional UV sources shows a significantly higher inactivation for pulsed light.

More than 7 log cycles of Aspergillus niger spore inactivation result with few pulsed light flashes. A variety of microorganisms including Bacillus subtilis are inactivated by using between 1 and 35 pulses of light with intensity ranging between about 1 and 12 J cm^{-2}. Greater inactivation is obtained when full spectrum light rather than glass-filtered spectrum light is used. Thus, the UV component of light is needed to inactivate microorganisms. Spores of Bacillus subtilis, Bacillus pumilus, Bacillus stearothermophilus and Aspergillus niger were inactivated completely from 6–8 logs of colony forming units (CFU) with 1–3 pulses. Bacillus subtilis, for example, is sterilized (99.999%) by about 42,600mW s cm^{-2} of UV while requiring a dose of only 4500mW s cm^{-2} under pulsed light. Microorganisms are inactivated by a combination of photochemical and photothermal mechanisms.

Light pulses apparently do not affect the nutrient retention in food, although a detailed study is not available. Pulsed light treatment costs are very favourable. Packaging materials compatible with this process will transmit light over the broad spectrum employed. LLDPE, LDPE, nylon, Aclar, HDPE and PP have all been used. Package geometry should not allow any shadowing on the product or the light exposure may not be sufficient.

Chemical Methods

Thermal destruction is by far the most common and efficient mode of sterilization. Dry or moist heat for sterilization is also not always feasible. Chemical methods use a wide variety of chemicals in the form of liquids and gases to disinfect and sterilize equipment and packaging materials.

Hydrogen Peroxide

Hydrogen peroxide is one of the most widely used sterilants used for sterilizing packaging materials. The first successful aseptic filling system for cartoning the aseptic Tetra Pak of 1961 used a combination of hydrogen peroxide and heat for the sterilization of the surface of container material. Many aseptic packaging systems use hydrogen peroxide at concentrations from 30 to 35% as a sterilant for packaging materials and other food contact surfaces followed by hot air (60–125 °C) to augment the sterilizing effect and to dissipate residual hydrogen peroxide. Sterilizing performance increases with both peroxide concentration and temperature. At ambient environmental temperatures, the activity of hydrogen peroxide is relatively slow; however, this can be increased considerably by raising the temperature to 85–90 °C and/or increasing concentration. Therefore for effective sporicidal action in food processing environment, treated with hydrogen peroxide (at 30%) is followed by application of hot air. Researchers found that four decimal reductions were obtained after suspension in 20% at 80 °C for 15 s. Researchers state that the challenge organisms of choice are Bacillus stearothermophilus for aseptic packaging systems that use heat sterilization and strains of Bacillus subtilis A or Bacillus subtilis var. globigii for filler surfaces that sterilize with hydrogen peroxide and heated air. The following relationship holds for destruction of spores of Bacillus stearothermophilus:

$$\log D = 17 - 1.3 * \log(\% H_2O_2)$$

where D is the decimal reduction time in minutes.

Spores resistant to hydrogen peroxide have been reviewed by Ito, and Stevenson and Shafer. Toledo et al. reported that dry spores of Bacillus subtilis var. globigii (NCIB 8058) were less resistant than the wet spores. Consistent with observations for heat and irradiation, dry spores are more resistant than wet spores. Since sterilization processes used on aseptic packaging systems for low-acid foods must have documentation for adequacy to inactivate pathogenic and spoilage microorganisms, a unit must be tested using an acceptably resistant organism before it can be placed in commercial production. In commercial aseptic packaging equipment which uses roll stock, the packaging material is immersed in hydrogen peroxide solution followed by heating to vaporize the peroxide before the packages are filled. Contact time with the solutions, which contain a wetting agent, is often less than 1 min. A large amount of the sterilizing liquid is removed mechanically, e.g. by rollers or air blasts, and the remainder is generally removed by drying with hot or sterile air or radiant heat. The nascent oxygen thus produced due to breakdown of peroxide reacts with the oxidizable cell components of the microorganisms and causes bactericidal effect.

Commercial packaging systems employ hydrogen peroxide at concentrations of 10–35% at either room or elevated temperature in the form of a spray or immersion bath. Hydrogen peroxide kills a wide variety of organisms and viruii. Both linear and non-linear relationships between log number of survivors and time of exposure to different concentrations of hydrogen peroxide have been reported and researcher's data show that 1 min in 30% hydrogen peroxide at ambient temperature produced only 2.5 decimal reductions of mould spores and bacterial spores were not affected at all. Thus, in current commercial practice, the microbicidal action is primarily due to the heated film of hydrogen peroxide at the start of the drying phase of the sterilization process. For this reason, wetting of the packaging material and the presence of a uniform film of liquid on the material surface are critical factors.

When sterilizing pre-formed containers, hydrogen peroxide is sprayed or atomized into the container. A measured amount of hydrogen peroxide is metred into each nozzle which delivers the solution into each container to ensure that a uniform film coats the inside surface of the package. For acid food products, relatively low microbial lethality required makes it easy to achieve the desired level of microbial inactivation and meet the residue tolerance level. However, for low-acid foods, a system must be carefully designed such that both criteria of commercial sterility and adequate sterilant removal are achieved.

Hydrogen peroxide-sterilized aseptic packaging systems that have been approved for commercial packaging of low acid foods have been proven to produce an acceptable level of residues. The Food and Drug administration of the USA has ruled that the level of hydrogen peroxide that may be present in product packaged in material sterilized by hydrogen peroxide must not be greater than 0.5 ppm. An improperly designed system may result in more incidence of non-compliance with regulations pertaining to residues. Prolonged contact between liquid hydrogen peroxide and the packaging material also increased the difficulty of removing residues.

Hydrogen peroxide is frequently used to sterilize the packaging material as it decomposes to a product which will neither taint the food nor render it toxic. It is reported that the traces of residual of hydrogen peroxide left in food is less than 0.25 ppm and normally this amount does not have any harmful effect when consumed. Residual hydrogen peroxide and that trapped in the package headspace at the time of sealing has an adverse effect on product stability particularly on ascorbic acid degradation in fruit juices. When the concentration of hydrogen peroxide at the time of packaging exceeded 0.1 μg/ml^{-1} a marked increase in ascorbic acid degradation in bottled orange juice and orange juice concentrates was observed by Toledo. Aseptic packaging of foods containing ascorbic acid in a system that uses hydrogen peroxide for sterilization of material is not desirable, unless some means is used to remove hydrogen peroxide vapours from the package head space and packaging material is completely free from hydrogen peroxide residues.

There are three commercial methods for applying hydrogen peroxide to packaging materials: dipping, spraying, and rinsing.

1. Dipping: The packaging material (i.e. plastic laminates with cardboard, films of thermoformable plastics and laminates) are taken from a reel and dipped into a bath of aqueous 30–33% hydrogen peroxide. Count reductions on the order of 4–5 logs are claimed. Wetting agents are added to ensure uniform wetting of the surfaces. Excess solution is removed by squeeze rolls or air jets after removal of the material from the bath, which leaves a thin film of solution that is then dried by the application of hot air. To increase the efficacy, especially in the case of dusty or slightly soiled material, prior treatment of the material with rotating brushes, sterile compressed air jets or ultrasound applied to the bath may be added.

2. Spraying: This is the method for preformed containers in which the hydrogen peroxide is sprayed in the form of small dispersed droplets into the container. Spraying does not result in a cohesive film due to the hydrophobic characteristics of plastics; rather only 30–40% of the inner surfaces of the container are covered and thereby it does not ensure good sterility of film. A conventional spray gives drops of over 30 mm diameters on the surface, and only 30–40% of the surface area is covered. An ultrasonic system can be used to give particle sizes of only 3 mm diameter, which will give an average surface cover of about 60%. The droplet size distribution, between 2 and 80 mm, and the distribution of droplets on the inner surfaces of the various geometric shapes of the containers is problematical. The efficacy is dependent upon the volume of solution sprayed; however, the larger the volume, the longer the drying time. The drying must be carried out with hot sterile air. Improved aerosol sprayers can limit the droplet size to 2–4 mm and increase the coverage to 60% while permitting reduced volume of peroxide solution to be applied; hence the drying time. This method is being replaced by the use of a mixture of hot air and vapourized peroxide. Sterilization by hydrogen peroxide vapour would be a cost-effective alternative as the least amount of hydrogen peroxide is used. Furthermore, the amount of hydrogen peroxide adsorbed on the treated surface from the vapour phase will be several orders of magnitude smaller than a liquid film. Therefore flushing the vapour-treated surface with low-temperature sterile air free of hydrogen peroxide vapours can effectively eliminate residues.

Low levels of hydrogen peroxide, which remain in the vapour phase in air at near ambient temperature has been shown to have sporicidal activity suitable for in-line sterilization of aseptic packaging materials. Air saturated with hydrogen peroxide vapour carry relatively low concentration of hydrogen peroxide (7.6 mg 1^{-1} at 70 °C) but can induce 6 decimal reductions of spores of Bacillus subtilis var. niger in 1.2 min. In comparison the same organism requires 1.2 min to achieve one decimal reduction in hot air alone at 150 °C. Since the amount of hydrogen peroxide vapourized from solution is less than that from water, the solution is enriched during prolonged operation and a system must be devised to add make-up water to maintain the desired concentration.

3. Rinsing: Prefabricated, intricately shaped containers for which the spraying process is unsuitable can be rinsed with an aqueous 30–33% H_2O_2 solution. For ambient temperature sterilization this can be combined with peracetic acid. The containers are drained, allowed to drip dry, and then completely dried by hot sterile air. Glass containers, metal cans, and blow-moulded plastic bottles are treated in this manner.

Ethylene Oxide

Implementation of ethylene oxide sterilization has made possible the use of sterile, low-cost, disposable thermoplastic devices for industrial application. It has been used to presterilize paperboard cartons and plastic packaging materials. However, no commercial system uses ethylene oxide as a sterilizing agent in aseptic packaging of low acid foods for ambient temperature storage and distribution. The vapour form is ammable and explosive and hence employed as a mixture with an inert gas such as dichlodi uoromethane. Researchers reported that 100% ethylene oxide has better microbicidal properties than an ethylene oxide inert gas mixture. An important consideration in bulk sterilization of preformed packaging materials is ethylene oxide's ability to permeate and contact all surfaces that need to be sterilized. The use of lower concentration of ethylene oxide is safer than use of 100% ethylene oxide from standpoint of potential for injury to personnel.

The D-value for Clostridium botulinum 62 A require a minimum of 72 min for a 12D process, thus making it impractical to use ethylene oxide for in-line sterilization of aseptic packaging materials in high-speed packaging lines. Ethylene oxide is effective against bacteria, moulds, yeasts, as well as viruses. The most probable activity is alkylation of nucleic acids and other cellular components. It is a particularly effective sterilizing agent because it rapidly penetrates packaging material. The ethylene oxide concentration, humidity and temperature influence the rate of sterilization. A clean object can be sterilized if treated for 5–8 h at 38 °C or 3–4 h at 54 °C when the relative humidity is maintained at 40–50% and the EtO concentration at 700 mg l^{-1}. Extensive aeration of the sterilized materials is necessary to remove residual EtO because it is so toxic. However, one disadvantage of it is its comparatively slow rate of action against microorganisms. Spores of Bacillussubtilis var. niger ATCC 9372 are used as indicators in dry heat and ethylene oxide sterilization.

Other Sterilization Methods

There are many other chemicals such as peracetic acid, beta propiolactone, alcohol, chlorine and its oxide and ozone etc. that have been suggested as having potential for use in sterilizing aseptic packaging materials. Peracetic acid is a peroxide of acetic acid and on decomposition it forms acetic acid and water. The low pH and oxidizing properties make it an excellent sporicidal agent. It has all the advantages of hydrogen peroxide and is unaffected by catalase and peroxidases. Peracetic acid is a liquid sterilant which is particularly effective against spores of aerobic and anaerobic bacteria and is effective at lower temperatures than hydrogen peroxide. In practice, it is used in a solution which also contains hydrogen peroxide.

The solution containing peracetic acid and hydrogen peroxide is effective against resistant bacterialspores even at 20 °C, for example a 1% solution will eliminate 10^{-7}–10^{-8} of most resistant spore stains in 5 min at 20 °C, and the most resistant strains in 60 min. The maximum usable temperature is 40 °C when the sterilization times are about 5 times shorter. In spite of its sporicidal properties, peracetic acid is not an approved sterilant for use on aseptic packaging materials. Its vapour is very pungent and irritating. Therefore a system that utilizes this compound must be airtight to prevent environmental release. Unlike hydrogen peroxide vapours for which a tolerance is allowed inside sealed packages, no such tolerance is allowed for peracetic acid and vapours in the headspace can cause disagreeable vinegar like off-flavour in some food products. In spite of these problems, this compound deserves attention as a possible sterilant for aseptic packaging materials because of its effectiveness as a sporicidal agent. The presence of small amounts of acetic acid may not be considered off-flavours. At ambient temperature and halogen concentration of 100 ppm, exposure times for effective sterilization would be on the order of hours. Under these conditions, it would be impractical to use these compounds as in-line sterilants in aseptic packaging. One major problem with the use of strong chlorine preparations for sterilization is that available chlorine reacts with organic material. Reaction between chlorine and the material being sterilized may produce off-flavours.

Beta propiolactone is bactericidal, sporicidal, fungicidal and virucidal. It lacks the penetrating power of the ethylene oxide, but it is considerably more active against microorganisms. Only 2–5 mg of beta propiolactone per litre is required for sterilization compared with 400–800 mg of ethylene oxide per litre. However, its poor penetration of materials and its cancer-inducing properties have restricted its use as a practical sterilizing agent.

Ethanol is the only industrially significant alcohol used as bactericide and fungicide but not for sporicide. Their mechanism of action appears to involve protein denaturation and membrane lipids dissolution. Ethanol liquid and vapour have been suggested. However, ethanol is effective only against vegetative cells and not against fungal conidia or bacterial spores therefore its use in packaging of foods is limited to extension of shelf life of packaged foods, which are normally stored under refrigeration. The two most popular alcohol germicides are ethanol and isopropanol, usually used in about 70–80% concentration.

Chlorine and its oxide and ozone are used mainly for water disinfection. Ozone can be used as a sterilizing agent because of its high oxidizing capability. Death of almost all microorganisms usually occurs within 30 min. Ozone was recommended recently as an alternative to chlorine and hydrogen peroxide. They found that hydrogen peroxide at 10,000-fold higher concentration was less effective than ozone against Bacillus spores. Antimicrobial power of ozone increases as temperature decreases below ambient. Comparatively low concentration is needed to eliminate large population of spores at ambient temperature in short time period and this makes ozone best suited for an industrial setting.

Antimicrobial Food Packaging

Food-borne microbial epidemics are driving a search for innovative ways to inhibit microbial growth in the foods while maintaining quality, freshness and safety. One option is to use packaging to provide an increased margin of safety and quality. The next generation of food packaging may include materials with antimicrobial properties. These packaging technologies could play a role in extending shelf-life of foods and reduce the risk from pathogens. Antimicrobial polymers may be useful in other food contact applications as well. Antimicrobial food packaging acts to reduce, inhibit or retard the growth of microorganisms that may be present in the packed food or packaging material itself.

Types of Antimicrobial Packaging

Addition of Packets/Pads Containing Antimicrobial Agents to Packages

The most successful commercial application of antimicrobial packaging has been sachets or packets that are enclosed loose or attached to the interior of a package. The forms predominated are oxygen and moisture absorbers & ethanol vapour generators. Oxygen and moisture absorbers are used primarily in bakery, pasta produce and meat packaging to prevent oxidation and water condensation. Although oxygen absorbers may not be intended to be antimicrobial, a reduction in oxygen inhibits the growth of aerobes, particularly molds. Moisture absorbers can reduce a_w, also indirectly affecting microbial growth. Ethanol vapour generators consist of ethanol encapsulated in carriers and enclosed in polymer packets. The ethanol permeates the selective barrier and is released in the headspace within the package. Since the amount of ethanol generated is relatively small and effective only in products with reduced water activity (a_w < 0.92), applications have been mainly to retard molds in bakery and dried fish products. Commercial examples include Ethicap, heat sealed packets containing microencapsulated ethanol in silicon dioxide powder, and Fretek, a paper wafer in which the centre layer is impregnated with ethanol in acetic acid and sandwiched between layers of polyolefin films. One of the drawbacks is the characteristic off flavour of ethanol.

Incorporation of Antimicrobial Agents Directly into Polymers

Incorporation of bioactive agents including antimicrobials into polymers has been commercially applied in drug and pesticide delivery, household goods, textiles, surgical implants and other biomedical devices. Research on the incorporation of antimicrobials into packaging for food applications has increased more than 100% in the past 5 years. GRAS, non-GRAS and 'natural' antimicrobials have been incorporated into paper, thermoplastics and thermosets, and have been tested against a variety of microbes including Listeria monocytogenes, pathogenic E. coli, and spoilage organisms including molds. Of all the antimicrobials, silver substituted zeolites are the most widely used as polymer additives for food applications, Sodium ions present in zeolites are substituted by silver ions, which are antimicrobial against a wide range of bacteria and molds. These substituted zeolites are incorporated into polymers like polyethylene, polypropylene, nylon and butadiene styrene at levels of 1-3%. Silver ions are taken up by microbial cells disrupting the cells' enzymatic activity. Commercial examples of silver substituted zeolites include Zeomic, Apacider, AgIon, Bactekiller and Novaron. Other compounds like antimicrobial enzymes such as lactoperoxidase and lactoferrin, antimicrobial peptides such as magainins, cecropins, defensins, natural phenols like hydroquinones and catechins, fatty acid esters, antioxidant phenolics, antibiotics and metals like copper and others may be useful.

The rationale for incorporating antimicrobials into the packaging is to prevent surface growth in foods were a large portion of spoilage and contamination occurs. For example, intact meat from healthy animals is essentially sterile and spoilage occurs primarily at the surface. This approach can reduce the addition of larger quantities of antimicrobials that are usually incorporated into the bulk of the food. The gradual release of an antimicrobial from a packaging film to the food surface may have an advantage over dipping and spraying. In the latter processes, antimicrobial activity may be rapidly lost due to inactivation of the antimicrobials by food components or dilution below active concentration due to migration into the bulk food matrix. Emulsifiers and fatty acids are known to interact with nisin reducing the bacteriocin's activity.

Antimicrobials may be incorporated into polymers in the melt or by solvent compounding. Thermal polymer processing methods such as extrusion and injection moulding may be used with thermally stable antimicrobials. Silver substituted zeolites, for example, can withstand very high temperatures (up to 800 °C) and therefore have been incorporated as a thin co-extruded layer with other polymers. For heat-sensitive antimicrobials like enzymes and volatile compounds, solvent compounding may be a more suitable method for their incorporation into polymers.

Antimicrobial packaging materials must contact the surface of the food if they are non-volatile, so the antimicrobial agents can diffuse to the surface, therefore, surface characteristics and diffusion kinetics become important. Antimicrobial release from the polymer has to be maintained at a minimum rate so that the surface concentration is above a critical inhibitory concentration. To achieve appropriate controlled release to the food surface, the use of multilayer films (control layer/matrix layer/barrier layer) has been proposed. The inner layer controls the rate of diffusion of the active substance while the matrix layer contains the active substance and the barrier layer prevents migration of the agent towards the outside of a package.

Coating or Adsorbing Antimicrobials to Polymer Surfaces

Early developments in antimicrobial packaging incorporated fungicides into waxes to coat fruits and vegetables and shrink films coated with quaternary ammonium salts to wrap potatoes. Other early developments included coating wax paper and cellulose casings with sorbic acid for wrapping sausages and cheeses.

Antimicrobials that cannot tolerate the temperatures used in polymer processing are often coated onto the material after forming or are added to cast films. Cast edible films, for example, have been used as carriers for antimicrobials and applied as coatings onto packaging materials and/or foods. Examples include nisin/methylcellulose coatings for polyethylene films and nisin/zein coatings for poultry. Proteins have an increased capacity for adsorption due to their amphiphilic structure. Nisin adsorbed onto silanized silica surfaces inhibited the growth of L. monocytogenes. Other examples include: adsorption of nisin on PE, EVA, PP, polyamide, PET, acrylics and PVC, pediocin containing milk-based powders adsorbed onto cellulose casings and barrier bags and nisin/EDTA/citric solutions coated onto PVC, nylon and LLDPE films.

Manipulating the solvents and/or polymer structures can enhance antimicrobial adsorption. Polyethylene- co-methacrylic acid films treated with sodium hydroxide and swollen with acetone showed an increased absorption and diffusion of benzoic and sorbic acids compared to non-treated films. These NaOH-treated films also had the highest inhibitory effect on molds. The explanation is that the higher polarity of NaOH-treated films enhanced the absorption of the antimicrobials. Binders such as polyamide resins have also been used to increase compatibility between polyolefins surfaces and bacteriocins. Glucose oxidase has been coated onto moisture proof fabric sheets by using polyvinyl alcohol, starch and casein as adhesives.

Immobilization of Antimicrobials by Ionic or Covalent Linkages to Polymers

This type of immobilization requires the presence of functional groups on both the antimicrobial and the polymer. Examples of antimicrobials with functional groups are peptides, enzymes, polyamines and organic acids. Examples of polymers used for food packaging that have functional groups are EVA, EMA, Ionomer, Nylon, PVDC etc. In addition to functional antimicrobials and polymer supports, immobilization may require the use of 'spacer' molecules that link the polymer surface to the bioactive agent. These spacers allow sufficient freedom of motion so the active portion of the agent can contact microorganisms on the food surface. Spacers that could potentially be used for food antimicrobial packaging include dextrans, polyethylene glycol, ethylenediamine and polyethyleneimine, due their low toxicity and common use in foods.

Ionic bonding of antimicrobials onto polymers allows slow release into the food. However, diffusion to the product is less of a concern when the antimicrobial is covalently bonded to the polymer unless conditions within the product promote reactions such as hydrolysis. This may occur for example, during the heating of a high acid food.

Lysozyme and chitinase, both active against Gram positive bacteria, have been covalently immobilized. But, the activity was too low to be practical for packaging commercial applications. Glucose oxidase catalyzes the reaction between glucose and oxygen yielding the antimicrobial hydrogen peroxide. This enzyme has been covalently bound onto insoluble supports that could be compatible with packaging materials. Betagalactosidase and glucose oxidase have been co-immobilized with the objective of producing hydrogen peroxide to activate lactoperoxidase in milk. Other antimicrobial enzymes that could potentially be covalently immobilized for packaging applications include lactoferrin, sulfhydril oxidase and bile-salt stimulated lipase. A major challenge, is the incorporation of substrates into the system as well as managing undesirable products from the reactions. For example, glucose oxidase needs glucose as a substrate, which could be provided by the food or added externally. Lactoperoxidase requires hydrogen peroxide and thiocyanate, commonly present in milk but not in many other foods. In both systems, hydrogen peroxide may raise toxicological concerns if amounts exceed the regulatory levels.

Use of Inherently Antimicrobial Polymers

Some polymers are inherently antimicrobial and have been used in films and coatings. Cationic polymers such as chitosan and poly-L-lysine promote cell adhesion since charged amines interact with negative charges on the cell membrane, causing leakage of intracellular constituents. Chitosan has been used as a coating and appears to protect fresh vegetables and fruits from fungal degradation. Although the antimicrobial effect is attributed to antifungal properties of chitosan, it may be that the chitosan acts as a barrier between the nutrients contained in the produce and microorganisms. In addition, chitosan-based antimicrobial films have been used to carry organic acids and spices. Calcium alginate films reduced the growth of the natural flora and coliform inoculum on beef, possibly due to the presence of calcium chloride. Bactericidal acrylic polymers made by co-polymerizing acrylic protonated amine co-monomer have been proposed as packaging materials for increased fruit and vegetable shelf life.

Physical modification of polymers has been investigated as means to render surfaces antimicrobial. For example, the antimicrobial potential of polyamide films treated with UV irradiation has been reported. Antimicrobial activity was presumably the result of an increase in amine concentration on the film's surface. Positively charged amine groups present in polymer surfaces may enhance cell adhesion but not necessarily death. It is possible that in the tests mentioned, simple adsorption occurred, masking the lack of antimicrobial activity of the laminated polymer surface. In a study on UV-treated nylon films it is seen that the surface amino groups were bactericidal, but that bacterial cells were adsorbed to the surface and diminished the effectiveness of the amine groups. In many cases, these studies are conducted in buffer. Addition of nutrients could potentially prevent cell membrane damage and bacterial recovery and/or inhibit the adhesion of the cells to the surface due to the interaction of salts and other cations with the surfaces.

Mode of Antimicrobial Function in the Package

Antimicrobial functions which are achieved by adding antimicrobial agents in the packaging system or using antimicrobial polymeric materials show generally three types of mode; (i) release, (ii) absorption and (iii) immobilisation. Release type allows the migration of antimicrobial agents into foods or headspace inside packages, and inhibits the growth of microorganisms. The antimicrobial agents can be either a solute or a gas. Absorption mode of antimicrobial system removes essential factors of microbial growth from the food systems and inhibits the growth of microorganisms. For example, the oxygen-absorbing system can prevent the growth of moulds inside packages. Immobilisation systems do not release antimicrobial agents but suppresses the growth of microorganisms at the contact surface. Immobilisation systems are less effective in the case of solid foods compared to the liquid foods because there is less possibility for contact between the antimicrobial package and the whole solid food products.

Applications for Food Preservation

For both migrating and non-migrating antimicrobial materials, intensive contact between the food product and packaging material is necessary and therefore potential food applications include especially vacuum or skin-packaged products, e.g. vacuum-packaged meat, fish, poultry or cheese. Until now only few antimicrobial methods have found applications as a food-packaging material. But, as research in the area continues, more systems are showing promise for commercial applications.

Meat Products

Studies have focused on establishing methods for coating LDPE film with methyl cellulose as a carrier for nisin. These films have significantly reduced the presence of Listeria monocytogenes in solutions and in vacuum packaged hot dogs. Use of chitosan to inhibit L. monocytogenes and chlorine dioxide sachets for the reduction of Salmonella on modified atmosphere-packaged fresh chicken breasts was also observed. Suitability of the antimicrobial films with incorporated lactic acid or sodium lactate, for packaging of raw chicken meat was studied. The chicken meat packaged in lactic acid/sodium lactate releasing films was storable at 4 °C for 7 days and retardation of microbial growth was by more than 2 log cycles.

Fruits, Vegetables and Cereals

Oranges when coated with an emulsion containing an essential oil (including basil essential oil) or a volatile compound (Citrus reticulata Blanco) had a longer shelf life than uncoated ones. Oranges treated with geraniol were rendered almost completely free from blue mold decay. Other treatments, in a decreasing order of effectiveness, were found to be mentha and basil essential oils. Basil oils can cause total inhibition of fungal development on maize kernels. The optimal dosage for protection of maize was 5% (v/v) with hexane as the solvent. In addition, no phytotoxic effect on germination and corn growth was detected with this oil. Fresh salad can be preserved by delivering essential oils to the product in the washing solution of the vegetables, or a food packaging film containing these oils can also be used.

Dairy Products

Effect of polyethylene films coated by commercial PVC lacquer with addition of nisin (5% w/w) and/or natamycin (10% w/w) on the growth of target bacteria, yeasts and moulds on the surface of soft cheese was observed. The cheese in contact with nisin/natamycin treated film was stored at temperature 4 °C and 23 °C for 23 days, released antimicrobial agents caused retardation of microorganism growth by more than 1 log cycle on the surface of cheese.

Modified Atmosphere Packaging

Modified Atmosphere Packaging (MAP) is well established in the food industry and continues to gain in importance. MAP means, simply put, that the natural ambient air in the package is replaced by a gas or gas mixture, often nitrogen and carbon dioxide. This packaging under a protective atmosphere preserves the quality of fresh produce over a longer period of time, prolongs shelf-life, and gives food producers access to a geographically larger market for perishable products. This is suitable for meat and sausage products, dairy products, bread, fruits and vegetables, fish or convenience products.

Modified atmospheres are not only used in packaging. They can feature as part of the production process, e.g. in the case of minced meat, or of storage and transport, for example of fruit and vegetables in halls or containers. The standards required by

Modified Atmosphere Packaging are comparatively high, and have to be controlled and monitored to ensure safety. Therefore, food manufacturers rely on modern MAP gas technology and various levels of quality assurance for maximum process safety.

Benefits of Modified Atmospheres

- Longer shelf life/higher quality: Food packaged under a protective atmosphere spoils much slower. Combined with continuous cooling, Modified Atmosphere Packaging can significantly extend the freshness and shelf life. This effect varies depending on the product type. However, a doubling of the shelf life is usually possible. Normally, MAP products keep a high quality over a longer period of time and arrive at the consumer in the best possible condition.
- Less waste: Longer durability is often associated with fewer problems during long distance shipment, and longer shelf life. As a result, waste disposal due to spoiled food can often be reduced.
- More sales opportunities: Because of the longer shelf life, Modified Atmosphere Packaging typically opens up new geographic markets to manufacturers. Particularly with perishable goods, longer shipment distances can be achieved. A global market can become a reality.
- Fewer preservatives: Packaging under a protective atmosphere extends the shelf life of food, meaning in many cases that the use of preservatives can be reduced or even completely eliminated. Consumers get products that do not contain artificial additives.
- Appealing package design: Next to functional aspects, the design of the packaging plays a significant role in the competition for consumers. The look-and-feel and the quality impression influence the purchasing behaviour. Modified Atmosphere Packaging is very well suited for the most appealing packaging design and presentation of the food product.

Limitations of Modified Atmospheres

- Comparatively high complexity: The modified atmosphere packaging process involves comparatively high requirements. Possible failures: incorrect gas composition or leaks due to faulty temperature or pressure distribution, contaminated or worn tools, seal contamination or defective material. However, with modern MAP technology and comprehensive quality assurance, the risks can be mastered.
- Relatively high cost: In addition to high-quality films, the consumption of gas and the personnel costs for quality control are particularly costly. However, these costs can be minimised with efficient use of resources.

- Influence on product quality: Unlike using preservatives, in most cases, the protective gases are not absorbed by the food and thus do not alter the nature or taste of the product. But there are exceptions to this rule. For example, an excessively high concentration of CO_2 can be absorbed by the food and make it sour. However, these effects can be avoided with adapted gas mixtures. The influence of very high oxygen concentration on the quality of meat is controversial. Modified atmospheres are supposed to make the meat chewier. Data to support this is, however, sparse.

Factors Influencing the Shelf Life of Food and the Influence of Modified Atmospheres

From the time that fruits and vegetables are harvested or animals are slaughtered, the spoilage process begins. This process is often accelerated the more the products are processed, such as cut fruit or minced meat. How long foods are durable, which means suitable for consumption, is very different and depends on various factors, e.g. the content of water and salt, pH value, hygiene conditions during production, storage conditions such as temperature or humidity, packaging. Depending on the characteristics and combinations of these factors, food products are differently sensitive to microbial or chemical/biochemical spoilage.

- Chemical and biochemical spoilage: Directly after harvesting of plant or slaughter of animal material, chemical processes begin to change the structure or quality. Sometimes this is useful, e.g. dry-aging of meat, which can be seen as maturation to improve quality. In principle, however, the quality of organic material decreases. For example, the oxidation of fats quickly leads to a rancidity of the product.
- Microbial spoilage: Microorganisms are a major threat to the shelf life and quality of food. On the one hand, they influence colour and smell, but they can also lead to health hazards and make the products uneatable. The source of the microorganisms is either the food itself or an impurity that cannot be completely excluded in the production and packaging process.

  The changes due to chemical/biochemical and microbial spoilage can be significantly slowed by MAP techniques together with cooling. Various gases and mixtures with different properties are used to slow the process of spoilage as much as possible.

Typical Gases for Modified Atmosphere Packaging

Carbon dioxide (CO_2) and nitrogen (N_2) are mainly used as protective gases in food packaging. Carbon monoxide (CO) or argon (Ar) is also common in some countries. Oxygen (O_2) is also used in some cases.

- Oxygen (O_2) essentially causes food to spoil due to oxidation and forms the ideal preconditions for aerobic microorganisms to grow. As a result, oxygen is frequently excluded from modified atmosphere packaging. In some cases – typically red meat – processing is deliberately carried out with high oxygen concentrations, in order to prevent the red colour from becoming ,pale, and to inhibit the growth of anaerobic organisms.
- Carbon dioxide (CO_2) is colourless, odourless and tasteless. It has an oxidation-inhibiting and growth-inhibiting effect on most aerobic bacteria and moulds. The gas is frequently used to increase the shelf life of food. The shelf life of packaged or stored food is normally longer, the higher the CO_2 content. Nevertheless, many products can become sour if the dosage is too high. In addition, the gas can diffuse out of the packaging or be absorbed by the product – and the packaging thereby collapses. The use of supporting or filling gases can slow down this effect.
- Nitrogen (N_2) is an inert gas and owing to its production process, is typically relatively high purity. It is usually used for displacing air, especially atmospheric oxygen, in food packaging. This prevents the oxidation of food and inhibits the growth of aerobic microorganisms. It is frequently used as a supporting or filling gas, as it diffuses very slowly through plastic films and hence remains longer in the packaging.
- Carbon monoxide (CO) is colourless, odourless and tasteless. Similar to oxygen, carbon monoxide is sometimes used to retain the red colour of, primarily, meat. The required concentrations are very low. In some countries, including the EU, the use of carbon monoxide for modified atmospheres is nonetheless prohibited in foods.
- Argon (Ar) is inert, colorless, odorless and tasteless. Owing to the similarity of its properties to those of nitrogen, argon can replace nitrogen in many applications. It is believed that certain enzyme activities are inhibited and argon slows metabolic reactions in some types of vegetables. Due to the marginal effects and the higher price compared to nitrogen, its use is rather rare.
- Hydrogen (H_2) and helium (He) feature in modified atmospheres in some applications. However, these gases are not used to extend shelf life. They are used as trace gases for some leak detection systems available on the market. The relatively small molecular size of the gases allows rapid escape through packaging leaks. Since these gases otherwise have no positive properties on the food products and are expensive and not easy to handle, their use is rare. The most common method for leak testing is the detection of CO_2 which is the core component in many MAP processes. If food is packaged under a protective atmosphere, this must be indicated on the label. In addition, according to EU Regulation 95/2/EC, the gases used must be listed

with their corresponding E-number. The E numbers for the most important gases are:

- Argon E 938.
- Helium E 939.
- Carbon dioxide E 290.
- Oxygen E 948.
- Nitrogen E 941.
- Hydrogen E 949.

Foodstuffs Suited to Modified Atmosphere Packaging

Modified Atmosphere Packaging is suitable for a wide range of food product. While traditionally mainly dairy products, meat products or bread were packaged under protective atmosphere, now MAP is more and more used for other foods like fish, coffee, fruit or vegetables. In addition, Modified Atmosphere Packaging is driven by the growing popularity of ready-made meals and convenience products.

Meat and Sausage Products

Meat and sausage products, especially raw meat, are very prone to spoiling due to microbial growth, on account of their high moisture and nutrient content. No matter whether beef, pork or poultry – spoilage begins from the moment of slaughter and especially after butchering. Besides high hygiene standards and permanent cooling, modified atmospheres can significantly extend the shelf life of meat and sausage products. CO_2 is the most important among the protective gases.

At concentrations above 20 %, CO_2 can considerably reduce microbial growth. In the case of red meat, there is also the risk of oxidation of the red colour pigments. The meat will lose its red colour, becoming grey and unappetising in appearance. This oxidation is especially prominent with beef. A high oxygen content in protective gas packaging can prevent oxidation. A low carbon monoxide content (approx. 0,5 %) can also help to retain the red colour of meat. However, the use of this gas is not allowed in the EU, for example. Poultry is especially sensitive to rapid spoilage and is therefore subject to higher requirements for permanent cooling. Here too, a modified atmosphere with CO_2 content will extend the shelf life. A high oxygen content is also used for poultry without skin so as to retain the colour of the meat. The CO_2 can partly be absorbed by the foods. To prevent the packaging from collapsing, nitrogen is used as a supporting gas. Sausage and meat products, e.g. marinated or smoked meat pieces, react very differently depending on the preparation. Longer shelf-lives can be afforded by the use protective gases right from the start. The CO_2 content should not be too high with these products, in order to prevent a sour taste.

Fish and Seafood Products

Fish and seafood are some of the most sensitive foods. They are at risk of rapidly declining in quality and spoiling even shortly after the catch. The reason for this lies in the neutral pH value as an ideal precondition for microorganisms as well as special enzymes that negatively affect taste and odour. Fish, which is rich in fatty acids, also becomes rancid quickly. The most important element for a longer shelf life is cooling close to 0° Celsius. Modified atmospheres with minimum 20 % CO_2 also retard the growth of bacteria. CO_2 components around 50 % are frequently used. Higher CO_2 concentrations can lead to undesirable side effects such as liquid loss or a sour taste. In the case of low-fat fish and shellfish, O_2 is also used in the packaging. This prevents a fading or loss of the colour, while at the same time serving as a growth inhibitor for some types of bacteria. When dealing with shellfish and crustaceans, special attention should be paid to ensuring a CO_2 content that is not too high. This can be discerned most clearly by a sour taste, while these products absorb CO_2 the most, as a result of which the packaging can collapse. Nitrogen as an inert supporting gas prevents this effect.

Dairy Products

Cheese is predominantly spoiled by microbial growth or rancidness. A continuous cooling chain essentially extends the shelf life of products. With hard cheese, there is a risk of mould formation upon contact with oxygen. As a result, vacuum packaging was frequently used in the past, even though these are awkward to open and can leave unattractive marks behind on the product at the same time. CO_2 effectively prevents mould formation, but does not otherwise affect the maturation of the cheese. Soft cheese can quickly become rancid. This problem can also be tackled with CO_2 modified atmospheres. However, as soft cheese absorbs CO_2 to a significantly higher extent, there is a risk of the packaging collapsing. A correspondingly lower CO_2 content should therefore be chosen. In the case of milk products such as yoghurt or cream, there is a risk of the products absorbing too much CO_2 and becoming sour. A lower CO_2 content should therefore be chosen.

Milk powder, above all for use in baby food, is a highly sensitive product. It is especially important to ensure that oxygen is displaced from the packaging in order to extend the shelf life. In practice, packaging is carried out in pure nitrogen with as low a residual oxygen content as possible.

Bread and Cake

With bread, cake and biscuits, the shelf life is primarily affected by potential mould formation. A high standard of hygiene during production and packaging can significantly minimise this risk. Packaging involving a modified atmosphere with CO_2 and without oxygen largely prevents the products from becoming mouldy and extends the shelf life. To prevent the packaging from collapsing owing to CO_2 absorption by the products, nitrogen is used as a supporting gas in many cases.

Fruit and Vegetables

Modified atmospheres in packaging make it possible to offer consumers fresh and untreated products – in other words succulently fresh fruit and vegetables – with a long shelf life. At the same time, fruit and vegetables are subject to very special requirements in regard to the nature of the packaging and atmosphere. This is because – in contrast to other food – fruit and vegetables continue breathing after the harvest and consequently require an oxygen content in the packaging.

Furthermore, the packaging film does not have to be fully tight. By taking the product's breathing and the permeability of the film, typically via micro-perforation, into account, the composition of carbon dioxide, nitrogen and low amounts of oxygen ideal for the product can be maintained. The term used here is an EMA (equilibrium modified atmosphere). The gas composition is individually adapted to the corresponding product. Thorough cleaning along with hygienic processing are the fundamental preconditions for long-lasting freshness. Modified atmospheres, in conjunction with corresponding cooling, can be used to extend the shelf life of fresh produce, while achieving an attractive packaging design at the point of sale.

Pasta and Ready-made Meals

The nature and composition of fresh pasta and, in particular, readymade meals are very different. Above all, multi-component products such as ready-made pizzas or sandwiches contain many different foods with differing shelf lives and spoilage properties. In the majority of cases, modified atmospheres can significantly extend the shelf life without using oxygen. Mixtures of CO_2 and nitrogen are used here. The concentration of the gases is oriented to the content of the product. If, for example, there is a risk that large volumes of CO_2 will be absorbed by the product, the nitrogen content should be chosen higher to prevent the packaging from collapsing.

Snacks and Nuts

Snack products, for example potato crisps or peanuts, primarily involve problems associated with the fat content of the food. There is a risk of oxidation, whereby the products can quickly become rancid if the packaging is not optimal. To extend the shelf life, it is therefore important to minimise the contact with oxygen. Modified atmospheres with 100 % nitrogen are frequently used. In this way, a premature spoilage can be prevented, while these atmospheres also provide protection from mechanical damage to sensitive products, e. g. potato crisps in conventional packets.

Wine

Gases or gas mixtures are often used to protect wine in the different phases of its production process and to retain the quality of the product. They are mainly used to avoid contact with oxygen and prevent microbial deterioration. The tank headspace is replaced with an inert gas or a gas mixture, for example of CO_2, N_2 or Ar. The composition of the gases is chosen according to the type of wine.

Coffee

As a dried product, coffee is relatively insensitive to spoilage by microorganisms. However, the risk of the fatty acids it contains oxidising and making the product rancid is greater. To prevent this, oxygen is excluded from coffee packaging. Instead, a modified atmosphere comprising pure nitrogen is frequently used in coffee sachets or capsules.

Examples of Gas Mixture Compositions

Product	O_2	CO_2	N_2
Raw red meat	70	23-30	0-10
Raw offal	80	20	0
Raw poultry with skin	0	30	70
Raw poultry without skin	70	20-30	0-10
Cooked meat and sausage products	0	20-30	70-80
Raw low-fat fish	20-30	40-60	20-40
Raw high-fat fish	0	40	60
Cooked/smoked fish	0	30-60	40-70
Shellfish and crustaceans	30	40	30
Hard cheese	0	30-100	0-70
Soft cheese	0	10-40	60-90
Sliced cheese	0	30-40	60-70
Cream cheese	0	100	0
Yoghurt	0	0-30	70-100
Milk powder	0	0-20	80-100
Crispy breads	0	50-100	0-50
Cakes, biscuits	0	50	50
Fruit / Vegetables, fresh	3-10	3-10	80-90
Vegetables, cooked	0	30	70
Ready-made meals	0	30-60	40-70
Pasta/Pizza	0	30-60	40-70
Sandwiches	0	30	70
Snacks/Crisps/Peanuts	0	0	100
Wine, white / Rosé	0	20	80
Wine, red	0	0	100
Coffee	0	0	100

Quality Control of Modified Atmosphere Packaging

Modified Atmosphere Packaging makes comparatively high demands on the packaging process, especially on the sealing process. Many sources of error can lead to leaks, usually micro-leaks. Right from the point of mixing the gases and introducing them into the package, maximum care is required. A faulty mixture or leaking packaging can have serious effects - from the loss of nutrients, taste, colour or structure to a bad smell or infestation with microorganisms. Depending on the product, health risks to the consumer cannot be completely eliminated.

Modified Atmosphere Packaging therefore requires modern high quality equipment and uncompromising standards of hygiene. But even when using best available technology, failures can't be completely avoided. So comprehensive quality assurance activities are essential. These can begin during the packaging process, using inline gas analysis, which constantly monitors the composition of the modified atmosphere. After packaging, the packages must be tested for the correct gas mixture and for leaks. Only with this level of rigour can it be ensured that the full benefit of Modified Atmosphere Packaging is achieved and that the customer receives a top quality product.

Gas Technology for Modified Atmosphere Packaging

Packaging Machines

There is no one special packaging machine for Modified Atmosphere Packaging. Various types of machines from several suppliers do the job. Hand vacuum chamber machines are the simplest type of MAP machines. They are operated manually and are suitable especially for smaller companies. Pre-formed bags are put into the chamber and filled with the product. After closing the chamber, the machine creates a vacuum and replaces the air with the modified atmosphere before the packaging is finally sealed.

For larger packaging volumes normally automatic packaging lines are used. So called thermoform-fill-seal machines are using packaging film from a roll. The film is heated inside the machine and formed onto trays which get filled with food product. The next steps are similar to the hand chamber machine but are done automatically. In a vacuum chamber the air is replaced by a gas mixture. The trays are then sealed. Tray-sealer machines function in a similar way. Main difference: The trays are not made inside the machine but are pre-formed and just sealed with a film.

Form-fill-seal or flow-pack machines are a further type of machine. Horizontal or vertical machines are available. These machines form a tube from a film and place the product inside. The air inside the tube is replaced by permanent flushing with modified atmosphere before the individual packs are sealed.

Gas Mixers and Meters

In the packaging process the air inside the package is replaced by a gas or a gas mixture. Pre-mixed modified atmospheres are available in different mixtures and under several brand names. Today, in most cases on-site gas mixers are used to create these gas mixtures. MAP gas mixers provide verified gas quality and safety in the packaging process – for germfree and long shelf life food. But above all they offer high flexibility to the user. At the push of a button different mixtures can be produced in the shortest time on one packaging line, depending on the requirements of the product. Some company offers gas mixing and metering systems for all packaging machines used in the food industry, no matter whether it's vacuum packaging, thermos forming, flow pack or chamber packaging machine. The gas mixing systems are adjusted to the specific product type and processes, and require only basic installation requirements.

Gas Analysers

Gas analysers are essential for quality control in the MAP process. The monitoring can be done as continuous analysis during the packaging process or batch sampling after the packaging process. For continuous analysis, a gas analyser module is integrated into the gas mixing system. The gas analyser monitors the correct composition of the gas mixture. Sample testing is part of the quality control system of almost every company working with modified atmospheres. Via a needle, a sample is taken from the package. High quality gas analysers use modern sensors, very precise and rapid and requiring a very low gas volume. They are therefore also suited to packs with very small headspace, and a very low volume of gas inside the package. All data is logged and can be archived for complete QA documentation.

Leak Detection

Modified atmospheres perform only if the protective gas remains inside the package. The package has to be fully leak tight. As a freshness guarantee to retailers and consumers, package leak detection can also provide competitive advantage. Leak testing prevents needless returns, loss of prestige, legal consequences and, worst case, loss of business. To optimise quality assurance, the user can choose between solutions for sample or in-line testing – based on CO_2 or a water bubble test. Package leak detection systems reliably detect even the smallest leak, and are easy to operate. Furthermore, all tests can be digitally logged and documented for customers.

Ambient Air Monitoring

Gas monitoring systems for ambient air protect employees and make the use of gases such as carbon dioxide safer. This is not toxic but accumulates unnoticed in closed rooms and replaces the oxygen in the air. A concentration of 0.3 percent carbon dioxide in ambient air can be a health hazard. The allowed maximum concentration in the workplace is 0.5 percent. At five percent, headaches and dizziness may occur; eight percent and more lead to unconsciousness or even death. The gas level warning unit permanently monitors the concentration of the respective gas in the ambient air, and activates an acoustic and visual alarm when individually definable limits are exceeded. For food and vegetables, controlled atmospheres are not just used in packaging but also for the control of ripening control, in special ripening chambers with the help of ethylene. By using gas analysers, the ambient atmosphere can be monitored.

Materials Used in Food Packaging

4

Food packaging materials are the materials that are used to enclose the food to protect it against physical, chemical or biological contamination. The various packaging materials can be classified into metal packaging materials, glass packaging materials, plastic packaging materials, and edible and bio based packaging materials. All these types of packaging materials are described in this chapter in an easy-to-understand manner.

Packaging Materials

The first packages served as containers, and their principal function was to hold food and water. They were probably taken directly from nature, such as leaves and shells. Later, containers were fashioned from natural materials: wooden logs, woven plant fibers, pouches made from animal skins. The next containers developed by early societies were clay pots, which date back to 6000 B.C. The first known pottery is from Syria, Mesopotamia, and Egypt. Besides being functional, clay bowls, vases, and other vessels were an artistic medium that today provide important clues regarding the culture and values of ancient peoples. Although no longer a significant packaging medium, clay still continues to have a major artistic value.

Paper based Packaging Products

The use of paper and paperboards for food packaging dates back to the 17th century with accelerated usage in the later part of the 19th century (Kirwan 2003). Paper and paperboard are sheet materials made from an interlaced network of cellulose fibers derived from wood by using sulfate and sulfite. The fibers are then pulped and/

or bleached and treated with chemicals such as slimicides and strengthening agents to produce the paper product. FDA regulates the additives used in paper and paperboard food packaging (21 CFR Part 176). Paper and paperboards are commonly used in corrugated boxes, milk cartons, folding cartons, bags and sacks, and wrapping paper. Tissue paper, paper plates, and cups are other examples of paper and paperboard products.

Paper

Plain paper is not used to protect foods for long periods of time because it has poor barrier properties and is not heat sealable. When used as primary packaging (that is, in contact with food), paper is almost always treated, coated, laminated, or impregnated with materials such as waxes, resins, or lacquers to improve functional and protective properties. The many different types of paper used in food packaging are as follows:

- Kraft paper—Produced by a sulfate treatment process, kraft paper is available in several forms: natural brown, unbleached, heavy duty, and bleached white. The natural kraft is the strongest of all paper and is commonly used for bags and wrapping. It is also used to package flour, sugar, and dried fruits and vegetables.

- Sulfite paper—Lighter and weaker than kraft paper, sulfite paper is glazed to improve its appearance and to increase its wet strength and oil resistance. It can be coated for higher print quality and is also used in laminates with plastic or foil. It is used to make small bags or wrappers for packaging biscuits and confectionary.

- Greaseproof paper—Greaseproof paper is made through a process known as beating, in which the cellulose fibers undergo a longer than normal hydration period that causes the fibers to break up and become gelatinous. These fine fibers then pack densely to provide a surface that is resistant to oils but not wet agents. Greaseproof paper is used to wrap snack foods, cookies, candy bars, and other oily foods, a use that is being replaced by plastic films.

- Glassine—Glassine is greaseproof paper taken to an extreme (further hydration) to produce a very dense sheet with a highly smooth and glossy finish. It is used as a liner for biscuits, cooking fats, fast foods, and baked goods.

- Parchment paper—Parchment paper is made from acid-treated pulp (passed through a sulfuric acid bath). The acid modifies the cellulose to make it smoother and impervious to water and oil, which adds some wet strength. It does not provide a good barrier to air and moisture, is not heat sealable, and is used to package fats such as butter and lard.

Paperboard

Paperboard is thicker than paper with a higher weight per unit area and often made in multiple layers. It is commonly used to make containers for shipping—such as boxes, cartons, and trays—and seldom used for direct food contact. The various types of paperboard are as follows:

- White board—Made from several thin layers of bleached chemical pulp, white board is typically used as the inner layer of a carton. White board may be coated with wax or laminated with polyethylene for heat sealability, and it is the only form of paperboard recommended for direct food contact.
- Solid board—Possessing strength and durability, solid board has multiple layers of bleached sulfate board. When laminated with polyethylene, it is used to create liquid cartons (known as milk board). Solid board is also used to package fruit juices and soft drinks.
- Chipboard—Chipboard is made from recycled paper and often contains blemishes and impurities from the original paper, which makes it unsuitable for direct contact with food, printing, and folding. It is often lined with white board to improve both appearance and strength. The least expensive form of paperboard, chipboard is used to make the outer layers of cartons for foods such as tea and cereals.
- Fiberboard—Fiberboard can be solid or corrugated. The solid type has an inner white board layer and outer kraft layer and provides good protection against impact and compression. When laminated with plastics or aluminum, solid fiberboard can improve barrier properties and is used to package dry products such as coffee and milk powder. The corrugated type, also known as corrugated board, is made with 2 layers of kraft paper with a central corrugating (or fluting) material. Fiberboard's resistance to impact abrasion and crushing damage makes it widely used for shipping bulk food and case packing of retail food products.

Paper Laminates

Paper laminates are coated or uncoated papers based on kraft and sulfite pulp. They can be laminated with plastic or aluminum to improve various properties. For example, paper can be laminated with polyethylene to make it heat sealable and to improve gas and moisture barrier properties. However, lamination substantially increases the cost of paper. Laminated paper is used to package dried products such as soups, herbs, and spices.

Carton

Corrugated boxes simply refer to what is commonly known as: Carton. Corrugated boxes are the ones many probably consider as 'Carton' as it produces the large shipping, shoe & storage boxes. What a lot of people do not realize is that corrugated boxes also come in different types depending on the durability and strength of the box. Identifying a certain corrugated material, however, is easy. How do you determine the material? Through its corrugated medium (also known as fluting). Identifying a corrugated material is easy. It consists of 3 layers of paper, an outside liner, an inside liner and a corrugated medium (also known as fluting). The corrugated medium that gives it strength and rigidity.

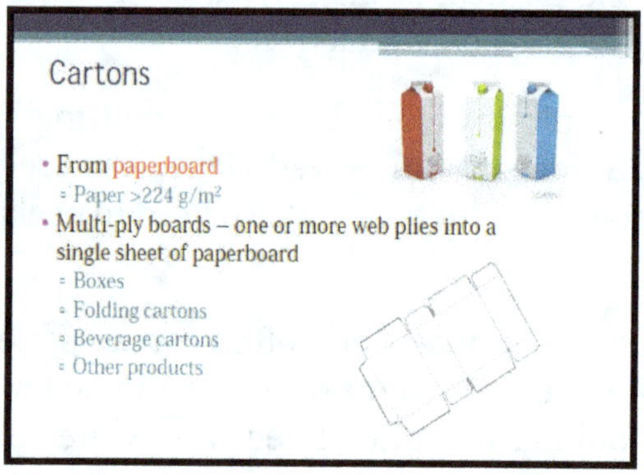

The main raw material that is used to construct the corrugated board is most recycled paper, made on large high-precision machinery known as corrugators. These type of boards can reused and recycled again and again as a source of pulp fiber. Corrugated boards are of different types, single faced, double faced (single wall), twin wall, and triple wall. They can be used to make packaging with different characteristics, performances, and strength. The board is cut and folded into different sizes and shapes to become corrugated packaging. Other applications of corrugated board packaging include retail packaging, pizza delivery boxes, small consumer goods packages, and so forth.

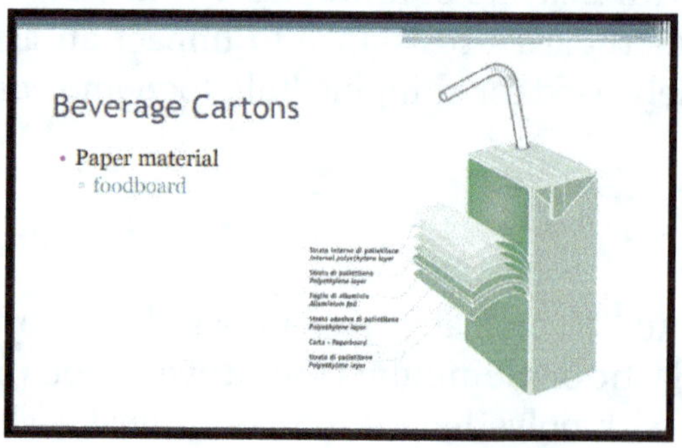

Chipboard Packaging

Chipboard packaging is used in industries such as electronic, medical, food, cosmetic, and beverage. A chipboard basically is a type of paperboard that is made out of reclaimed paper stock. It can be easily cut, folded, and formed. It is a cost-effective packing option for the products.

It comes in various densities and strength is determined by how high the density of the material is. If one want images to be directly printed onto the chipboard, one can treat the chipboard with bleach sulfate, and with CCNB (Clay Coated News Back) which makes the material even more durable.

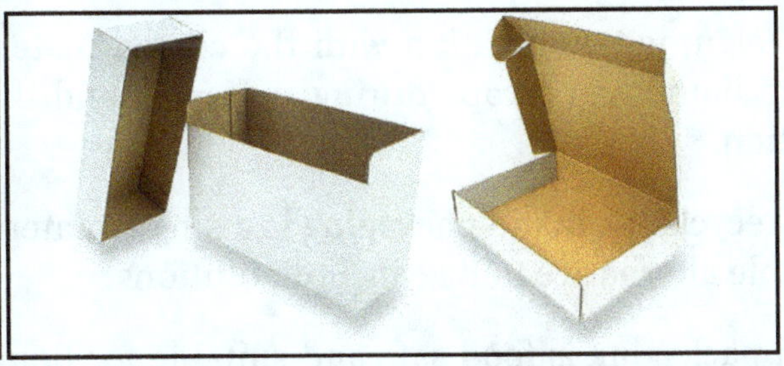

One cannot use chipboard packaging if the business deals with heavy items, since the chipboard sheet is a lightweight material, made for many grocery items such as cereal, crackers, tissue boxes, and so forth. Also, if the storage environment is heavy with moisture, chipboard can easily weaken, and result in discoloration and expansion.

Rigid Boxes

A rigid box is made out of highly condensed paperboard that is 4 times thicker than the paperboard used in the construction of a standard folding carton. The easiest real-world example of rigid boxes are the boxes that hold Apple's iPhones and iPads, which are 2 piece setup rigid boxes.

Compared to paperboard and corrugated boxes, rigid boxes are definitely among the most expensive box styles. The rigid boxes usually do not require dies that are expensive or massive machinery and are often hand-made. Their non-collapsible nature also gives them a higher volume during shipping, which easily incurs higher shipping fees.

These boxes are commonly used in merchandising cosmetics, jewelry, technology, and high-end luxury couture. It is easy to incorporate features such as platforms, windows, lids, hinges, compartments, domes, and embossing in a rigid box.

Liquid Packaging Board

Liquid tight packaging is designed with a special folding technique to ensure it is absolutely leak-proof and uses a recyclable coating system (RCS), which offers the highest barrier performance.

The design, in combination with the coated material, acts as a safeguards to ensure that no liquid will escape during transport and that your product arrives in optimal condition.

100% recyclable, liquid packaging is a single material solution using no plastics and is a reliable alternative to Bag-in-Box solutions.

Liquid packaging is food safe and suitable for liquid products such as cream concentrates or fruit pulp.

Features

- Bespoke to your exact size and performance requirements.
- Available in a wide range of corrugated cardboard and solid board grades.
- Uses a recyclable coating system for optimum liquid protection.
- Suitable for food contact.

- Leak-proof designs using a special folding technique.
- High quality print available to communicate your brand messages or product details.

Benefits

- Reduction in transit damage due to strength and stability of packaging.
- Environmentally friendly - Made from a renewable resource.
- Easy disposal - 100% recyclable in regular cardboard waste stream.
- Reduction in storage space and logistic costs as products can be delivered flat.
- Easy to use as no machines required to erect.

Sustainable Barrier Coatings

Liquid paperboard is providing a key market application for barrier coating materials that has a reduced environmental impact.

Several biopolymers have already been commercially trialled in coatings for paperboard. The most promising polymers identified were polyhydroxybutyrate (PHB) and polylactic acid (PLA). These two materials offer the best results, in terms of a consistent, thick and continuous layer, for extrusion coating, while other materials are better suited to dispersion and solvent casting.

Smaller Pack Sizes

Social changes are affecting packaging demand in unexpected ways. There is an increasing need for greater convenience and a growing demand for ready-to-drink and on-the-go foods and beverages. The result is a growing demand for smaller pack sizes in numerous food and beverage sectors.

Smaller packs consume more liquid paperboard to pack the same given quantity of product, leading to an increase in raw material consumption. Furthermore, these smaller packs consist of a greater surface area of material that requires more inks, adhesives and other consumables, stimulating value-added for converters.

Competing Materials

Competition from glass packaging and fluctuations in the price of aluminium are having a degree of influence on both liquid packaging board and multi-substrate flexible pouches.

Liquid packaging board faces competition from specific packaging formats and the latest of these is the clear retortable plastic can. This is likely to have a limited impact on the liquid paperboard market, but it does pose a threat to the retort carton, which has shown significant opportunities.

Recyclibility

Governments and beverage brand chains are increasingly pushing for recyclable hot beverage cups. A recent innovation in mid-2016 has seen Starbucks trial a fully recyclable cup as part of its efforts to reduce waste sent to landfill, as well as to improve its environmental profile.

Sustainability

A broad trend across the packaging industry is to minimize the environmental impact of packaging, which is stimulating both technical innovations and new business models across the value chain.

This interest includes consumers – especially millennials – brand owners, converters and regulators and spans a number of different means to realize the goal of furthering the use of sustainable, recyclable and compostable packaging into the mainstream.

Reduction of packaging waste can be achieved by various means, including lightweighting and by utilizing materials that are themselves sustainable, as in renewable and recyclable. This has been perceived to be an advantage of liquid paperboard, but new understandings of how polymer coatings inhibit this process, and the arrival of competing formats optimized for sustainability is threatening this.

An example of this competition is Coca-Cola's Plant Bottle concept. This is produced using up to 30% plant-based materials (biopolymers) including stems, barks and fruit skins, and Coca-Cola aims to produce all plastic bottles solely from recycled and bio-derived feedstocks , by 2020.

Healthy Eating Trends

In efforts to reduce welfare costs, various governments are endorsing healthier eating lifestyles and promoting healthier habits, which are being adopted by many consumers. The EU has targeted a 16% reduction in salt for processed foods, while the UK reduced the average consumption of salt by around 5% across 2005–2011.

Consumers have become more sensitive to prices, which has become evident in the strongly competitive food retail industry. In Western Europe, private labels are increasingly popular and are becoming a brand in their own right. The initial advantage that traditional brands had over private labels was the perceived quality differential, which still exists to differing extents in some countries, but is most evident in Italy. However, demand is changing, largely due to the poor economic climate. In addition, the improved quality branding of private labels has helped to put them on par with their branded equivalents.

Molded Pulp

Molded pulp is a type of product made from recycled paper that offers the benefits of cardboard or foam packaging without the drawbacks of non-sustainable materials. Sustainable packaging offers many benefits over traditional packing materials. Pulp products are created entirely from recycled paper, making it one of the most sustainable and environmentally friendly packaging solutions. This means that fewer trees are destroyed in creating more paper, and the paper that is used does not go into landfills as trash.

Manufacturing process consists of several steps:

- Recycled paper is collected from various sources and stored onsite until it is made into pulp moulded packaging.
- The recycled paper is sent through a pulping system, turning it into fiber pulp, the basic ingredient in our molded pulp tray products.
- Recycled water is added from a storage tank, turning the pulp into a slurry of paper and water.
- The pulp and water are dried through a vacuum system to extract excess water. The excess water is then pumped back to the recycled water tank. In this way, we reuse our water constantly and reduce the need for more water.

- The pulp material for the molded tray is formed and shaped into a unique configuration and dried completely to hold its shape. Our oven drying process uses heat exchangers. These reduce energy-use and makes our process more efficient.

Once the molded pulp has been processed and turned into convenient storage trays, it is shipped to customers for use as inner support shipping packaging. This molded pulp container can then be reused or recycled, continuing the cycle of sustainability.

Molds decide the shapes of final products, which are featured with elegant appearance, good stiffness, light weight and are easy to be stored and transported. Most important, raw material of these pulp molded products have a wide, cheap resource, and is easily available. Reed, straw, bamboo and various plant fibers can be utilized effectively. After application, all paper tableware can be 100% recycled because of their good biodegradability.

Due to excellent plasticity, any shape can be molded by a pulp moulding machine, all you have to do is just change dies.

Fast Food Tableware

Pulp molded tableware can be applied to dishware, bowl, fast food tray, cup carrier and more. These molded fiber products are featured with elegant appearance, good stiffness, light weight and are easy to be stored and transported.

They are water & oil proof, and can be froze by refrigerator or heated up by micro-wave oven, which is suitable for eating habits of modern people and fast food processing. According to a study, one ton of paper pulp can produce 40 thousands of 600ml standard dishware.

Egg Tray/Carton

Due to porous material, unique egg-shaped curved structure, better ventilation, freshness preservation and good cushioning, molded pulp egg tray/carton is especially suitable for packaging of high volume egg transportation, no matter chicken, duck, goose or quail egg.

Statistic shows that egg broken rate reclined to 2% from 8%~10% by using egg tray instead of traditional packaging method. Logo or other characters can be printed on egg tray/carton for exhibit impression. Just choose the right type egg try machine for your egg tray business.

Fruit Tray

Paper pulp can be molded into the shape of fruit, and used as fruit tray.

Molded Pulp Packaging of peach, pear, apple, orange, pineapple, tomato, and fruit for export, can avoid damage due to collision, distribute respiratory heat of fruit, absorb evaporated water, suppress ethylene density, prevent fruit from going bad and prolong freshness. All these actions are irreplaceable than other packaging material.

Egg Carton

Egg Cartons are the cartons used for carrying eggs.

Its functions are:

- Protection function.

  Egg trays can protect the egg from the external shock. It can also protect the eggs from damaging and going bad due to illumination and dampness so that the loss can be reduced.

- Convenient transportation.

  The purpose of producing eggs is to let them get into market, which is a process of product circulation. In this process, eggs will undergo countless times of transportation, handling, storage, which requires the egg packing must be able to adapt to the process. Egg trays can satisfy this purpose.

- Increasing sale.

  In today's market, great changes have taken place in the way of people's purchasing. With the improvement of people's material standard, more and more people begin to pursue a higher quality of life. The packaging quality of the products has a direct impact on sales.The egg is more typical. If we make the egg packaging exquisite and unique, it will increase the sales of eggs.

Metal Packaging Materials

Metal packaging is found everywhere in modern daily life due to its versatility and universal consumer approval. Its sustainability credentials combined with the outstanding characteristics listed below make it an ideal packaging solution that meets the changing demands of our society in the 21st century:

The advantages of the metal packaging:

- Keeps the content safe for longer.
- Prevents product waste.

- Keeps the packed nutrients clean from impurities.
- Extends the shelf life of the packed products.
- Ensures the protection form light, oxygen and bacteria – the perfect solution for modern lifestyle.
- The products remain fresh until the opening of the can.

Metal packaging with its extended shelf life provides protection against light, oxygen and bacteria and is an excellent solution for modern lifestyles. The can ensures products stay fresh until consumed contributing to value for money for consumers.

Metal continues to have immense potential for the development of innovative packaging solutions. The applications are unlimited, ranging from packs for chewing gum, microwaveable ready meals to aerosols and do-it-yourself (DIY) products.

Metal can solutions are continually being improved and developed to meet customer and consumer needs.

Hermetically closed, the metal packaging contributes to the complete preservation of the food from the moment of its closing.

Metal packaging allow the transportation and storage of harmful to people and environment materials.

Today, metal is used for a vast range of products and packaging solutions, yet designers continue to seek new opportunities to delight consumers. Cans come in many shapes, sizes and decorations for:

- Food.
- Beverages.
- Lifestyle and luxury products.
- Personal hygiene.
- Household products.
- DIY.
- Industrial.

Convenience

Convenience remains a key driver in packaging development for consumer packaged goods. Changes in family structures, flexible working and commuting hours, resulted in changing of the consumption habits. Smaller household units, more working women and mothers and busier lifestyles, the packaging industry has been constantly evolving to satisfy these changing needs.

Metal packaging enhances product safety by being both tamper proof as well as virtually unbreakable. Its barrier properties, which keep the goodness in and impurities out ensure that the content safely reaches the consumer, free from contaminants.

Metal packaging satisfies new demands from consumers via easy opening, re-sealable packaging and through manageable portions which allow consumers to open only what they need. Designs for metal cans which can be used in microwave ovens extend the convenience of metal packaging for ready meals.

Canned foods are safe to eat for many years – ideal for today's meal or tomorrow's emergency. Cans do not need refrigeration or freezing. They provide total protection against light, oxygen and, being hermetically sealed, also protect food to keep the nutrients locked inside. Products are also protected against moister, dust, rodents and other contaminants.

Logistics

Cans are typically 35% lighter than 20 years ago, but they are still unbreakable.

High-speed filling, at up to 1,500 cans a minute, saves significant amounts of energy compared with other packaging solutions.

Most can-making facilities are close to or next to filling lines, ensuring efficient and flexible delivery.

Stackable and tapered cans cater for efficient long-distance supply. Empty cans can be stacked over eight meters high, making the best use of warehouse space.

Embossing

Embossing and de-bossing technologies enable can makers to produce innovate can designs.

Embossing is a technology which produces decorations or reliefs with a raised profile (from inside out) while de-bossing produces decorations with a depressed profile (outside-in).

The advantage of de-bossing is that existing external dimensions are maintained and therefore there is no need to change transportation and pallet space.

The former metal is registered with printed graphics for maximum visual and tactile impact. Images and graphics, logos, tactile warning and brand identification features can all be enhanced.

Recycling

Growing concern for the environment has led to increased consumer interest in packaging materials and their recyclability.

Metal packaging is 100% recyclable, infinitely, without loss of its inherent properties. It is a permanently available resource that is widely recycled globally and one of the most recycled packaging materials in Europe.

Steel and aluminum are among the most abundant recourses in the world although recycled materials significantly reduce the impact of processing on the environment.

Metal has the best recycling and recovery rates of all key competing packaging materials and is improving year on year:

- Recycling rates in 2010 were 72% for steel cans and 64% for aluminium drinks cans in EU.
- Metal can be recycled an infinite number of times without loss of its essential properties.
- Metal packaging collection for recycling is simple and cost efficient: cans are retrieved easily by any one of the multiple collection systems used across Europe.
- Recycling reduces energy consumption and CO_2 emissions significantly.

Precise engineered features such as beading patterns and end panel designs have enabled light-weighting of the can by up to 35% over the last 20 years adding to its environmental credentials. Weight reduction will continue to be a feature of metal packaging in the years to come.

Metal packaging ranges from tin biscuit containers and aluminum to steel beverage cans. Metal is useful because it's durable, doesn't cost too much and is non-toxic, making it highly suitable for storing food. However, this kind of packaging also has a few disadvantages. Problems with metal packaging vary according to the specific type of metal used.

Corrosion

Some types of metal packaging, such as steel, are vulnerable to the effects of corrosion, which can cause the metal to deteriorate. Corrosion takes place as the metal begins to transform back into its original state; for example, steel turns back into the iron ore it came from. Corrosion is caused by oxidation, brought about when the metal is exposed to air and water. One example of corrosion is rust, which occurs on steel packaging and causes it to flake away. Metal packaging is typically coated in other materials, such as chromium, to prevent corrosion from occurring.

Can't See Contents

Metal packaging may keep a container's contents secure and fresh, but it does pose a disadvantage in that it is not transparent, and so consumers can't see into the packaging to check the contents or to further inspect a potential purchase. This limits the uses for metal packaging within the retail sector, since other packaging materials—such as plastic—are better in some situations. For example, the plastic used in a blister pack used to store nails allows consumers to check the size and type of nails inside the packaging, which wouldn't be possible with metal packaging.

Storage Issues

Tin is often used for certain types of containers, including those for biscuits. Becuase the metal packaging isn't easily bent or squashed by hand, the containers are difficult to store effectively, both during and after use. On the other hand, a paper or plastic container might be easier to fold up or squash and tuck away in a cupboard or other storage facility.

Aluminum and Acidity

Aluminum is another common choice for metal packaging. While aluminum is impervious to corrosion when used to store food products, it does have an issue with acidic foods such as rhubarb and tomatoes. These foods are especially acidic and can be affected by aluminum if the metal is used to store them. The result of using aluminum packaging for these foods is that the food will end up tasting of aluminum.

Metal Packaging Products

Aluminium Foil

Aluminum foil is made by rolling pure aluminum metal into very thin sheets, followed by annealing to achieve dead-folding properties (a crease or fold made in the film will stay in place), which allows it to be folded tightly. Moreover, aluminum foil is available in a wide range of thicknesses, with thinner foils used to wrap food and thicker foils used for trays. Like all aluminum packaging, foil provides an excellent barrier to moisture, air, odors, light, and microorganisms. It is inert to acidic foods and does not require lacquer or other protection. Although aluminum is easily recyclable, foils cannot be made from recycled aluminum without pinhole formation in the thin sheets.

Here are some notable benefits of using aluminium in food packaging.

- Helps in storing food for long periods.

  Because of its brilliant impermeability to water vapour and water, aluminium foils help to store food for prolonged time periods. Moreover, because of this wonderful property, aluminium helps to maintain the freshness of foods over long time periods.

- Provides barrier against heat and light.

  Because of its excellent ability to act as a barrier against heat and light, heat being one of the main reasons for food decay, aluminium is widely used in food packaging. This is one of the foremost properties that makes aluminium the most preferred element for packaging and storage of foods.

- Packaging is easy.

  A rather conspicuous benefit of using aluminium foils for food packaging is that it is easy to pack foods in an aluminium foil. All you need to do is wrap your food with the foil and make sure the wrapping does not squeeze the food packed inside.

- Prevents germs and bacteria.

  It isn't just heat that is responsible for deterioration in the quality of food. Aluminium foils prevent foods from coming in contact with bacteria and germs. It is highly resistant to germs and helps keep the food safe from the influence of harmful bacteria.

Aluminum Can

Aluminum cans provide long-term food quality preservation benefits. Aluminum cans deliver 100 percent protection against oxygen, light, moisture and other contaminants. They do not rust, are resistant to corrosion and provide one of the longest shelf lives of any type of packaging. Aluminum-based food canning has an unparalleled safety record. Tamper-resistant and tamper-evident packaging provides consumers with peace of mind that their products have been safely prepared and delivered. A vast variety of products are packaged using aluminum in addition to food and beverages: aerosol products, paint and thousands of other items in the consumer products market.

Beverage Can Technology Continues to Improve

In the past half-century, aluminum beverage can manufacturers have lightened the package by reducing the gauge required to fabricate both the cans and their ends. The first generation of aluminum cans weighed approximately 3 ounces per unit. Today's cans weigh less than half an ounce. Aluminum cans bring packaging benefits as well. They are easily formed, resist corrosion and will not rust. Cans made from aluminum easily support the carbonation pressure required to package soda and withstand pressures of up to 90 pounds per square inch. Believe it or not, four six packs can support a 2-ton vehicle.

Looking Forward: Aluminum Can Market Growth

Many new specialty drinks, particularly energy drinks, are delivered in 8.2-ounce cans. This move was propelled by the wildly successful Red Bull™ brand of energy drink, introduced in the U.S. market in 1997. Since then, the market has swelled with more than 30 new brands following Red Bull's™ lead. The microbrew beer canning market is also growing rapidly. Craft beer packaged in aluminum cans maintains a higher quality. The liquid is not affected by light or exposed to oxygen, which can negatively impact taste. Cans are also portable and infinitely recyclable—they can go where glass can't. Brewers are now positioning the aluminum can as a "mini-keg" that delivers draft-quality fresh beer to the glass.

History of Aluminum Beverage Cans

The modern aluminum beverage can traces its origins to 1959, when Coors introduced the first all-aluminum, seamless, two-piece beverage container. Recycling was instituted at the same time (Coors paid 1 cent for each can returned to the brewery). Aluminum cans made inroads into the soft drink market in 1964, when Royal Crown Cola released both its RC Cola and Diet Rite beverages in two-piece, 12-ounce aluminum containers. In their first year on the market, 1 million cases of soda were packaged using aluminum cans. In addition to being lighter in weight than their steel predecessors, aluminum cans provide a superior surface on which to print text and graphics. This capability increased the opportunity to establish and promote shelf presence and brand awareness.

Important Facts

- The most sustainable beverage container.

  As the most valuable package in the bin, aluminum cans are, by far, the most recycled beverage container. The average can contains 70 percent recycled metal.

- Shipping efficiency: the weight advantage.

  Aluminum cans are lightweight and easily stacked. This provides storage and shipping efficiencies and limits overall transportation carbon emissions through logistics and supply chains.

- True closed loop recycling.

Aluminum cans are recycled over and over again in a true "closed loop" recycling process. Glass and plastic are typically "down-cycled" into products like carpet fiber or landfill liner.

- New markets continue to develop.

  Nearly 500 craft beer brewers use aluminum to can more than 1,700 different beers. Protection from light and oxygen are two key benefits in addition to the unparalleled sustainability of aluminum packaging.

Metallised Film

Metallized PET foils are plastic foils which are only evaporated with aluminium powder in a high vacuum. For this process, it is used pure aluminium in powder form with an aluminium content of at least 99,98%. The films produced this way obtain a metallic shine and thereby achieve a special optical effect.

In addition to the optical effect, the aluminum vaporization serves in particular to improve its barrier effects against vapours, gases and aromas in comparison to non-metallized plastic film. Overall, the metallized PET films are less gas-tight than "real" aluminum foil, but more tear resistant, more flexible and lighter. Also, they are less sensitive to corrosion and not as winkle-prone as pure aluminium foil.

PET-met films are made of only 1% of the aluminum that would have been needed for an aluminum foil. Therefore, PET-met is especially interesting as a cost-effective and more environmentally friendly alternative to aluminum.

Metallized PET films are an ideal solution for all applications that require the preservation of flavor and freshness. It is suitable for many applications in the packaging industry as well as for applications in the technical and industrial sectors.

PET films have particular advantages where the higher gas tightness of the aluminum foil is not required.

Metallized PET film can be printed like aluminum foil, and it results in a variety of optical variants. The gloss level of metallized PET film is much higher than for aluminum foil, so with metallized PET film you can achieve a stronger mirror effect. This is also used when printing – for example, without a white primer, yellow colour shimmers gold.

Therefore, PET met films offer an ideal opportunity to combine certain protective and barrier functions with an attractive appearance – even if the protection of the food is not quite as high as with real aluminum foil and the shelf life may need to be shortened.

In combination with other materials, for example, PET met foils are ideal for snacks packaging, thermoformable container covers such as yogurt cups or refill bags for cosmetic products. Coffee and pet kibbles remain dry in those PET packaging and retain their full aroma and flavor for a long time.

In the particular case of acidic, alkaline or salt-containing foods, metallized PET films are a perfect substitute for aluminium foils, as a longer exposure or a large contact area of aluminium with those materials can release potentially harmful emission of aluminium into the edible content.

Insulation systems, for example for insulation against heat radiation in buildings or against heat conduction in vacuum insulation panels, can also be realized with metallized PET films. In addition, PET met films can be used for decorative adhesive tapes. Labels made from metallized PET-film are highly resistant to heat and chemicals, which ensures a long-time durability.

Difference between Aluminium Foil and Metallised Film

One of the major differences between aluminum foil and metallised film is the price, making it very tempting to stick with the cheaper option. However, the difference between semi-permeability and total protection can be significant depending on what you're packaging.

Differences in the permeability of polyethylene, metallised polyethylene and aluminum foil.

	Moisture (g/m². Day)	Oxygen (mL/m².day)	UV Light (%transmittance)
PET film, 12.7	31	465	91
Metallised PET	0.8	1.2	5
Aluminum foil 6	0	0	0

If you plan to store perishable food products and powdered substances in ziplock style bags, we recommend using the semi-permeable metallised film bags as they provide a durable barrier that allows some moisture and gas to pass through, preventing the food products from becoming rancid. For liquid based products such as juices, oils and lotions, we would recommend the aluminum foil bags.

- You can see through PET-met foils, not through aluminum foils.
- If high gas tightness or opacity is important, better use aluminum foil.
- For tomatoes, better use PET-met film instead of aluminium foil.

Beverage Can

The stay-tab opening mechanism characteristic of drinking cans.

A beverage can (or drinks can) is a can manufactured to hold a single serving of a beverage. In the United States, the can is most often made of aluminum (almost entirely), but cans made in Europe and Asia are an alloy of approximately 55 percent steel and 45 percent aluminum. Aluminum is a widely available, affordable, lightweight metal that is easy to shape. Also, it is far more cost-effective to recycle aluminum than to extract it from its ores.

Fabrication Process

Modern cans are generally produced through a mechanical cold forming process that starts with punching a flat blank from very stiff cold-rolled sheet. This sheet is typically alloy 3104-H19 or 3004-H19, which is aluminum with about one percent manganese and one percent magnesium to give it strength and formability. The flat blank is first formed into a cup about three inches in diameter. This cup is then pushed through a different forming process called "ironing," which forms the can. The bottom of the can is also shaped at this time. The malleable metal deforms into the shape of an open-top can. With the sophisticated technology of the dies and forming machines, the side of the can is significantly thinner than either the top and bottom areas, where stiffness is required. One can-making production line can turn out up to 2400 cans per minute.

Plain lids are stamped out from a coil of aluminum, typically alloy 5182-H49, and are transferred to another press that converts them to easy-open ends. The conversion press forms an integral rivet button in the lid and scores the opening, while concurrently forming the tabs in another die from a separate strip of aluminum. The tab is pushed over the button, which is then flattened to form the rivet that attaches the tab to the lid.

Finally, the top rim of the can is trimmed and pressed inward or "necked" to form a taper conical where the can will later be filled and the lid (usually made of an aluminum alloy with magnesium) attached.

Problems

One problem with the current design is that the top edge of the can may collect dust or dirt in transit, if the can is not packaged in a completely sealed box. Some marketers have experimented with putting a separate foil lid on can tops, and shipping cans in cardboard 12 or 24 pack cases.

Many consumers find the taste of a drink from a can to be different from fountain drinks and those from plastic or glass bottles. In addition, some people believe that aluminum leaching into the fluid contained inside can be dangerous to the drinker's health. The exact role (if any) of aluminum in Alzheimer's disease is still being researched and debated, though the scientific consensus is that aluminum plays no role in development of the disease.

Aluminum cans contain an internal coating to protect the aluminum from the contents. If the internal coating fails, the contents will create a hole and the can will leak in a matter of days. There is some difference in taste, especially noticeable in beer, presumably due to traces of the processing oils used in making the can.

Recycling

In many parts of the world, a deposit can be recovered by turning in empty plastic, glass, and aluminum containers. Unlike glass and plastic, scrap metal dealers often purchase aluminum cans in bulk, even when deposits are not offered. Aluminum is one of the most cost-effective materials to recycle. When recycled without other metals being mixed in, the can/lid combination is perfect for producing new stock for the main part of the can. The loss of magnesium during melting is compensated by the high magnesium content of the lid. Also, the refining of ores such as bauxite into aluminum requires large amounts of electricity, making recycling cheaper than smelting.

Advantantages of Cans

- Save energy: Reusing recycled metals saves as much as 95% of the energy needed to make cans from virgin ores.
- Recyclable: Beverage cans are fully and infinitely recyclable without any quality loss, again and again and again and again.

- Permanent material: The material in cans is only used, not consumed. Because they are infinitely recyclable, metals are a permanent resource.
- Quickly chilled: Beverage cans chill quickly and feel extrafresh to the touch.
- Unbreakable: Unbreakable beverage cans are ideal for large events.
- Material thickness: The sides of today's beverage cans are only 0.065 mm thick – as thin as a human hair. Thanks to ongoing research and development, it is now possible to manufacture cans with far less material than before.
- Light-proof: Beverage cans are absolutely light-proof, protecting the quality of light-sensitive beverages such as beer.
- Stackable: The flat ends and characteristic shape of beverage cans means that a truck carrying cans is able to transport twice as much liquid as a truck loaded with bottles.
- Hermetic seal: Being absolutely airtight, beverage cans keep oxygen out and fizz in, allowing beverages to stay fresh for longer.
- Lightweight: Light and convenient, beverage cans are great for refreshments on the way.
- Fresh: The characteristic sound of a can opening is a unique indicator that the drink inside is absolutely fresh.

Glass Packaging Materials

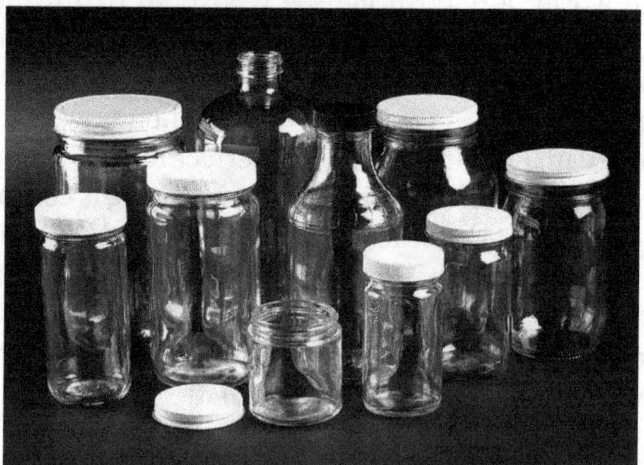

The two main types of glass container used in food packaging are bottles, which have narrow necks, and jars and pots, which have wide openings. Glass closures are not common today, but were once popular as screw action stoppers with rubber washers and sprung metal fittings for pressurized bottles, e.g. for carbonated beverages, and vacuumized jars, e.g. for heat preserved fruits and vegetables. Ground glass friction fitting stoppers were used for storage jars, e.g. for confectionery.

Glass Containers Market Sectors for Foods and Drinks

A wide range of foods is packed in glass containers. Examples are as follows: instant coffee, dry mixes, spices, processed baby foods, dairy products, sugar preserves, spreads, syrups, processed fruit, vegetables, fish and meat products, mustards and condiments etc. Glass bottles are widely used for beers, wines, spirits, liqueurs, soft drinks and mineral waters. Within these categories of food and drinks, the products range from dry powders and granules to liquids, some of which are carbonated and packed under pressure, and products which are heat sterilized.

The glass package has a modern profile with distinct advantages, including:

- Quality image – consumer research by brand owners has consistently indicated that consumers attach a high quality perception to glass packaged products and they are prepared to pay a premium for them, for specific products such as spirits and liqueurs.
- Transparency – it is a distinct advantage for the purchaser to be able to see the product in many cases, e.g. processed fruit and vegetables.
- Surface texture – whilst most glass is produced with a smooth surface, other possibilities also exist, for example, for an overall roughened ice-like effect or specific surface designs on the surface, such as text or coats of arms. These effects emanate from the moulding but subsequent acid etch treatment is another option.
- Color – as indicated, a range of colors are possible based on choice of raw materials. Facilities exist for producing smaller quantities of nonmainstream colors, e.g. Stolzle's feeder color system.
- Decorative possibilities, including ceramic printing, powder coating, coloured and plain printed plastic sleeving and a range of labelling options.
- Impermeability – for all practical purposes in connection with the packaging of food, glass is impermeable.
- Chemical integrity – glass is chemically resistant to all food products, both liquid and solid. It is odourless.
- Design potential – distinctive shapes are often used to enhance product and brand recognition.

- Heat processable – glass is thermally stable, which makes it suitable for the hot-filling and the in-container heat sterilization and pasteurization of food products.
- Microwaveable – glass is open to microwave penetration and food can be reheated in the container. Removal of the closures is recommended, as a safety measure, before heating commences, although the closure can be left loosely applied to prevent splashing in the microwave oven. Developments are in hand to ensure that the closure releases even when not initially slackened.
- Tamper evident – glass is resistant to penetration by syringes. Container closures can be readily tamper-evidenced by the application of shrinkable plastic sleeves or in-built tamper evident bands. Glass can quite readily accept preformed metal and roll-on metal closures, which also provide enhanced tamper evidence.
- Ease of opening – the rigidity of the container offers improved ease of opening and reduces the risk of closure misalignment compared with plastic containers, although it is recognized that vacuum packed food products can be difficult to open. Technology in the development of lubricants in closure seals, improved application of glass surface treatments together with improved control of filling and retorting all combine to reduce the difficulty of closure removal. However, it is essential in order to maintain shelf life that sufficient closure torque is retained, to ensure vacuum retention with no closure back-off during processing and distribution.
- UV protection – amber glass offers UV protection to the product and, in some cases, green glass can offer partial UV protection.
- Strength – although glass is a brittle material glass containers have high top load strength making them easy to handle during filling and distribution. Whilst the weight factor of glass is unfavourable compared with plastics, considerable savings are to be made in warehousing and distribution costs. Glass containers can withstand high top loading with minimal secondary packaging. Glass is an elastic material and will absorb energy, up to a point, on impact. Impact resistance is improved by an even distribution of glass during container manufacture and subsequent treatment.
- Hygiene – glass surfaces are easily wetted and dried during washing and cleaning prior to filling.
- Environmental benefits – glass containers are returnable, reusable and recyclable. Significant savings in container weight have been achieved by technical advances in design, manufacture and handling.

Glass Pack Integrity and Product Compatibility

- Safety.

  Migration studies on glass have shown it to be an inert material as regards its application to packaged foods and, from a health and hygiene viewpoint, it is regarded as an optimal material for containing food and drinks.

- Product compatibility.

  Glass containers are noted for the fact that they enable liquid and solid foods to be stored for long periods of time without adverse effects on the quality or flavour of the product.

- Consumer acceptability.

  Market research has indicated that consumers attach a high quality perception to glass packaged products. Recent findings of a report on consumer perceptions carried out by The Design Engine, on behalf of Rockware Glass, concluded that there are five key and largely exclusive benefits for food packaging in glass, namely:

  1. Aesthetic appeal.
  2. Quality perception.
  3. Preferred taste.
  4. Product visibility and associated appetite appeal.
  5. Resealability.

Glass and Glass Container Manufacture

Melting

Glass is melted in a furnace at temperatures of around 1350°C (2462° F) and is homogenized in the melting process, producing a bubble-free liquid. The molten glass is then allowed to flow through a temperature controlled channel (fore hearth) to the forming machine, where it arrives via the feeder at the correct temperature to suit the container to be produced. For general containers, suitable for foods and carbonated beverages, this would be in the region of 1100°C (2012° F).

Container Forming

In the feeder the molten glass is extruded through an orifice of known diameter at a predetermined rate and is cropped into a solid cylindrical shape. The cylinder of glass is known in the trade as a *gob* and is equivalent in weight to the container to be produced. The gob is allowed to free-fall through a series of deflectors into the forming machine, also known as the IS or individual section machine, where it enters the parison. The parison com- prises a neck finish mould and a parison mould, mounted in an inverted position. The parison is formed by either pressing or blowing the gob to the shape of the parison mould. The parison is then reinverted, placed into the final mould and blown out to the shape of the final mould, from where it emerges at a temperature of approximately 650°C (1200° F). A container is said to have been produced by either the press and blow or blow and blow process.

In general terms, the press and blow process is used for jars and the blow and blow process for bottles. An alternative, for lightweight bottles, is the narrow neck press and blow process.

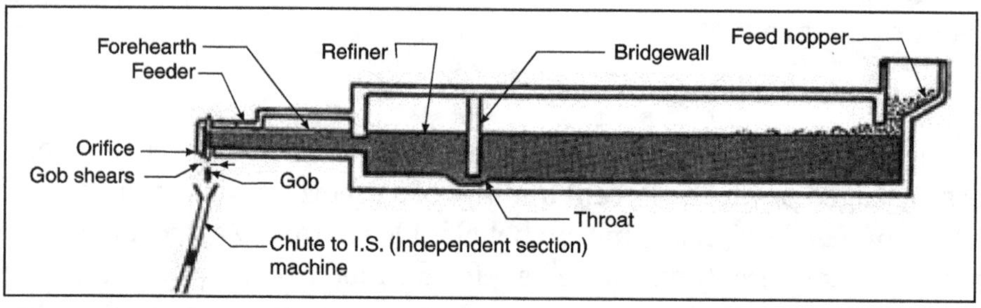

Figure: The feeder – molten glass is extruded through the orifice at a predetermined rate and is cropped into a solid cylinder known as a gob.

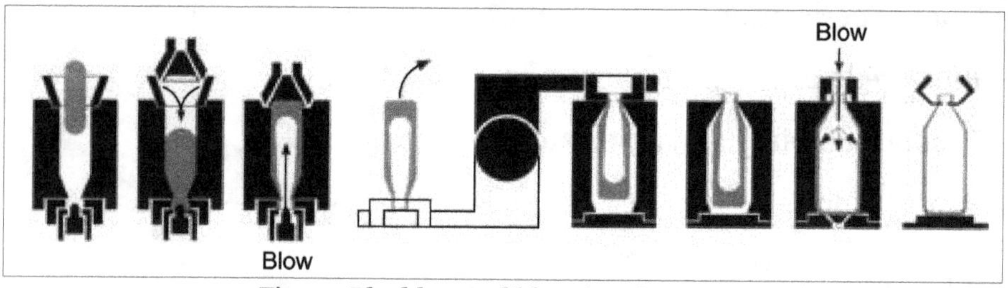

Figure: The blow and blow forming process.

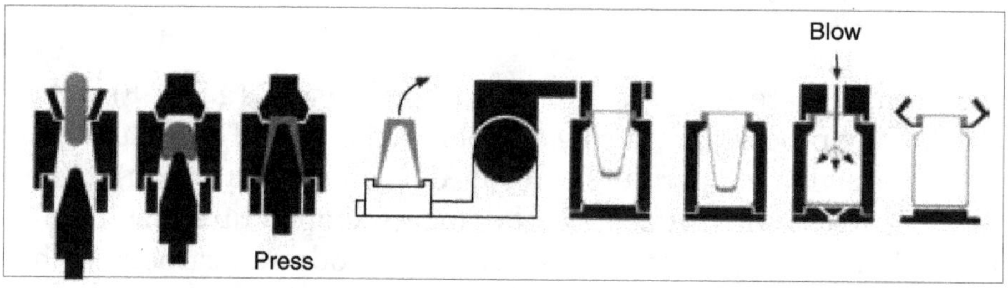

Figure: The wide mouth press and blow forming process.

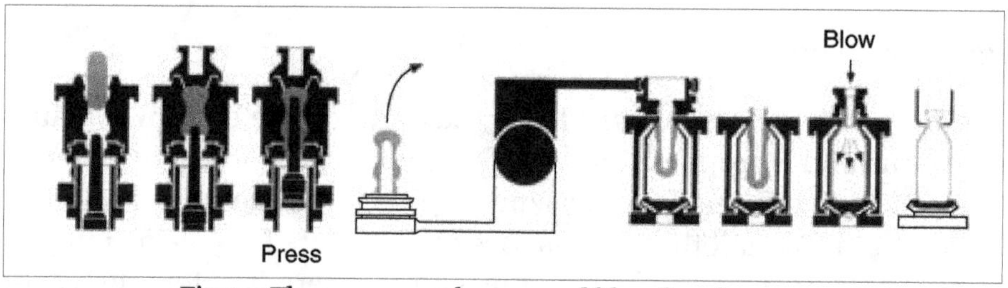

Figure: The narrow neck press and blow forming process.

The press and blow process is generally best suited to produce jars with a neck finish size of ≥35 mm (≥1.25″); the other two processes are more suited to produce bottles with a neck finish size of ≤35 mm (≤1.25″).

The narrow neck press and blow process offers better control of the glass distribution than the blow and blow process, allowing weight savings in the region of 30% to be made.

Design Parameters

One of the design parameters to be borne in mind when looking at the functionality of a glass container is that the tilt angle for a wide-mouthed jar should be ≥22° and that for a bottle ≥16°. These parameters are indicative of the least degree of stability that the container can withstand.

Surface Treatments

Once formed, surface treatment is applied to the container in two stages: hot end and cold end treatment, respectively.

Hot End Treatment

The purpose of hot end surface treatment is to prevent surface damage whilst the bottle is still hot and to help maintain the strength of the container.

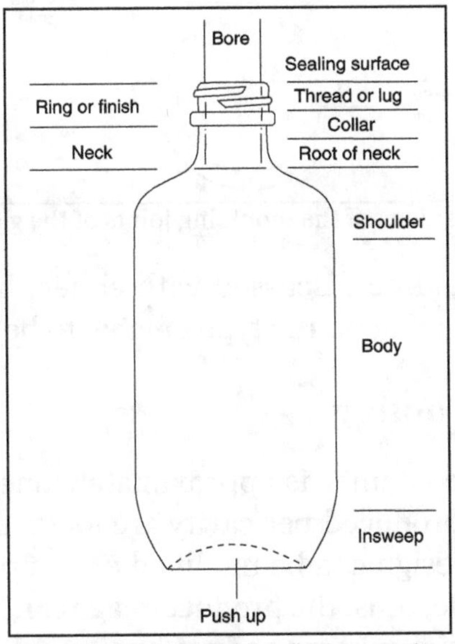

Figure: The parts of a glass container.

The most common coating material deposited is tin oxide, although derivatives of titanium are also used. This treatment tends to generate high friction surfaces; to overcome this problem, a lubricant is added.

Cold End Treatment

The second surface treatment is applied once the container has been annealed. Annealing is a process which reduces the residual strain in the container that has been introduced in the forming process. The purpose of the cold end treatment is to create a lubricated surface that does not break down under the influence of pressure or water, and aids the flow of containers through a high speed filling line. Application is by aqueous spray or vapor, care being taken to prevent entry of the spray into the container, the most commonly used lubricants being derivatives of polyester waxes or polyethylene. The surface tension resulting from this treatment can be measured by using Dynes indicating pens.

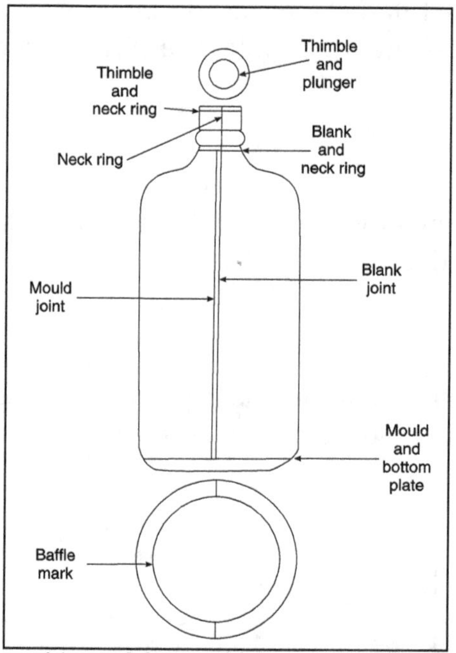

Figure: The positions of the moulding joints of the glass container.

Labelling compatibility should be discussed with either the adhesive supplier or the adhesive label supplier depending on the type of label to be used.

Low-cost Production Tooling

The tooling cost for a glass container is approximately one-fifth that of a plastic container. Whilst the numbers produced per cavity are lower than for plastic, this can be advantageous, because the design can be modified or completely revamped in a much shorter time-span than plastic; thus, the product image can be updated and the product marketability kept alive. The numbers produced per mould cavity vary depending on the number of production runs required, the complexity of the shape and the embossing detail. In general, 750 000 pieces can be produced from a complex mould and 1 000 000 pieces from a mould of a simple round shape. There can be upwards of 20 moulds per production set.

Container Inspection and Quality

As with packaging in general, quality assurance is needed to ensure that consumer safety, brand owner's needs and efficiency in handling, packing, distribution and merchandising are achieved.

Quality assurance needs are defined and incorporated into the specification of the glass container at the design stage and by, consistency in manufacture, thereby meeting the needs of packing, distribution and use. Quality control, on the other hand, comprises the procedures, including on-line inspection, sampling and test methods used to control the process and assess conformity with the specification.

The techniques used can broadly be defined as chemical, physical and visual. Chemical testing by spectrophotometry, flame photometry and X-ray fluorescence is used to check raw materials and the finished glass. Small changes in the proportions and purity of raw materials can have a significant effect on processing and physical properties.

Physical tests include checking dimensional tolerances, tests for color, impact strength, thermal shock resistance and internal pressure strength. Visual tests check for defects that can be seen.

The list of possible visually observable defects is quite long and though most of them are comparatively rare, it is essential that production be checked by planned procedures. The categories of these defects comprise various types of cracks, glass strands, foreign bodies and process material contamination from the process environment, misshapes and surface marks of various kinds.

Defects are classified as being:

- Critical, e.g. defects which endanger the consumer or prevent use in packing;
- Major, e.g. defects which seriously affect efficiency in packing;
- Minor, e.g. defects that relate to appearance even though the container is functionally satisfactory.

Visual inspection on manufacturing and packing lines is assisted today by automatic monitoring systems, as shown in figure below, where this is appropriate. Systems are available for container sidewall inspection using multiple cameras that detect opaque and transparent surface defects. Infrared cameras can be used in a system to examine containers directly after formation. On the packing line, claims have been made that foreign bodies can be detected in glass containers running at speeds up to 60 000 capped beer bottles per hour.

An X-ray system such as that from Heimann Systems Corp. is designed for the automatic inspection of products packed in jars.

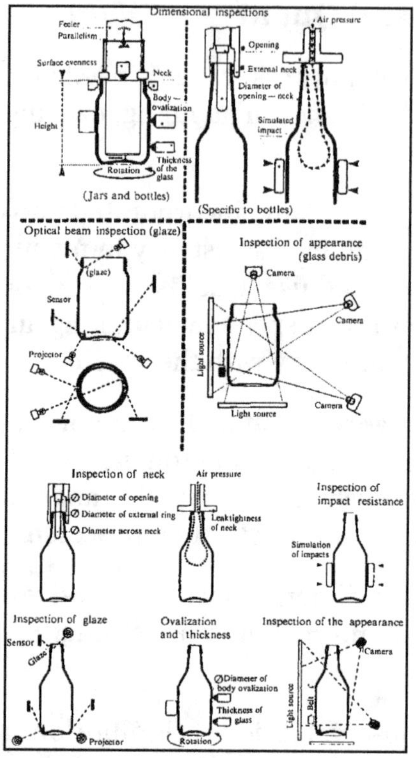

Figure: Principles of production line inspections.

The system can detect foreign materials such as ferrous and stainless metals, glass particles, stone, bone and plastic materials. This equipment runs at 100 m min^{-1}.

Closure Selection

Closures for glass packaging containers are usually metal or plastic, though cork is still widely used for wines and spirits. Effecting a seal is achieved either by a tight fitting plug, a screw threaded cap applied with torque in one of several ways or a metal cap applied with pressure and edge crimping. Hermetic or airtight sealing can be achieved by heat sealing a flexible barrier material to the glass usually with an overcap for protection and subsequent reclosing during use. The aluminium foil cap applied to a milk bottle is one of the simplest forms of closure.

All these closures are applied to what is known as the finish of the container. This may seem an odd name for the part of the container which is formed first but in fact this name goes back to the time of blowing and forming glass containers by hand when the rim was the last part to be formed and therefore called the finish.

Four key dimensions determine the finish as shown in figure. Industry- wide standards for these dimensions have been agreed upon. The contour of glass threads are round, and closures, both metal and plastic, with symmetrical threads will fit the appropriate containers.

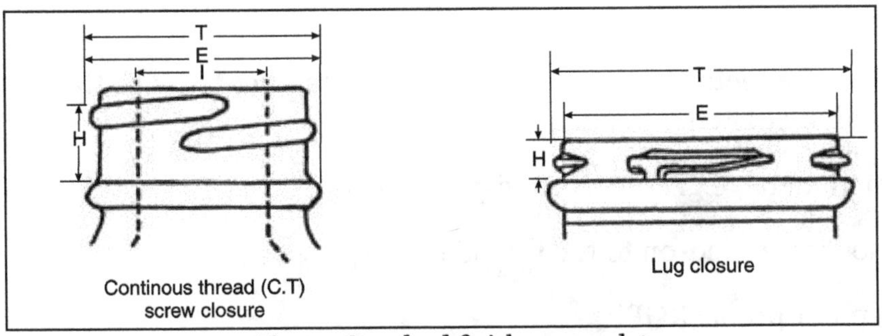

Figure: Standard finish nomenclature
I: Diameter at smallest opening inside finish, T: Thread diameter measured across the threads, E: Thread root diameter, H: Top of finish to top of bead or to intersection with bottle shoulder on beadless designs.

Careful choice of closure is essential. Too large a closure can create leakage due to the force generated upon it either from internal gas pressure or from heating during processing of the product. Too small a closure may well intro- duce an interference fit between the minimum through bore on the glass container and the filler tube. The types of closure available fall into three main categories:

1. Normal seal
2. Vacuum seal
3. Pressure seal.

Normal Seals

Normal seals, that is those used for non-vacuum/non-pressure filled products, comprise composite closures of plastic/foil, for products such as coffee, milk powders, powder and granular products in general and for mustards, milk and yoghurts.

Glass lends itself both to induction and conduction sealing without prior treatment of the glass finish, but is considered suitable only for dry powders and products, such as peanut butter and chocolate spreads, which do not require a further heating process.

Crimped seals using foils, used for instance on milk containers, have been used for a long time, and are cost effective.

Vacuum Seals

Vacuum seals are metal closures with a composite liner to seal onto the glass rim. They can be pressed or twisted into place, at which time a vacuum is created by flushing the headspace with steam. They lend themselves quite readily to in-bottle pasteurization and retort sterilization and sizes range from 28 to 82 mm. For beverages, sizes are usually in the 28–40 mm range.

Pressure Seals

Pressure seals can be metal or plastic with a composite liner to make the seal, and can either be pressed or twisted into place. They include:

- Preformed metal, e.g. crown or twist crown.
- Metal closures rolled-on to the thread of the glass.
- Roll-on pilfer proof (ROPP).
- Preformed plastic screwed into position with or without a tamper evidence band.

Selecting the correct glass finish to suit the closure to be used is essential. Advice on suitability should be sought from both the closure and the glass manufacturers before the final choice is made.

Thermal Processing of Glass Packaged Foods

Glass containers lend themselves to in-bottle sterilization and pasteurization for both hot and cold filled products. Subject to the headspace volume conditions being maintained and thermal shock ground rules being observed, no problems will be experienced.

In general terms, hot-fill products filled at 85°C and then cooled will require a minimum headspace of 5%, whilst a cold filled product requiring sterilization at 121°C will require a 6% minimum head space. In all cases the recommendations of the closure supplier should be obtained before preparing the design brief. It should be noted that the thermal shock of glass containers is twice as high when cooling down as when warming up. To avoid thermal shock, cool down differentials should not exceed 40°C and warm up differentials should not exceed 65°C.

Internal pressure resistance: A well-designed glass container can withstand an internal pressure of up to 10 bar, although the norm required rarely exceeds 5 bar. It is also capable of withstanding internal vacuum conditions and filling of thick concentrates, with steam-flushing of the head- space to produce the initial vacuum requirements for the closure seal.

Resealability: Preformed metal, rolled-on metal and preformed plastic closures can all be readily applied to the neck finish of a glass container. Prise off crown closures offer no reseal, whilst the twist-off crown satisfies reseal performance within reason.

Plastic Sleeving and Decorating Possibilities

Glass containers can accept a wide range of decorative formats, i.e. labelling, silk screen printing with ceramic inks, plastic sleeving, acid etching, organic and inorganic color coating and embossing. The rigidity of the container offers a good presentation surface for decorating, which is not subject to distortion from internal pressure or internal vacuum.

When plastic sleeving the container, it is essential to test the sleeving film under in-bottle pasteurization temperatures to ensure that no secondary movement of the sleeve occurs. Care should also be taken not to exceed the stretch limits of the film by ensuring the maximum and minimum diameters of the area to be sleeved do not fall outside the stretch ratios of the film specification.

Sleeving also offers fragment retention properties, should the container become damaged in use.

Strength in Theory and Practice

The theoretical strength of glass can be calculated and it is extremely high. In practice, the strength is much lower due to surface blemishes, such as micro- cracks, which are vulnerable stress points for impacts such as occur during handling and on packing lines. Work has therefore been concentrated on:

- Improving the surface to reduce defects.
- Improving surface coatings in manufacture.
- Avoiding stress during manufacture and use.

Major investigations of packing line performance, noting all breakages, and using techniques such as high speed video, can lead to improvements in performance by eliminating stress points.

Broken bottles can be reconstructed and thereby demonstrate the type of impact which caused the failure, such as whether it occurred at a slow or fast rate or whether it was caused by an external, or internal, pressure related fault. The strength of a glass container is also dependent on shape and thickness. The inter-relationship can be subjected to computer modeling as a design-aid to:

- Identify vulnerable features in a proposed design.
- Calculate the effect of modifying the design.
- Simulate the effect of lightweighting by reducing thickness.

Specific tests can be carried out on containers to check:

- Vertical crushing.
- Internal pressure.
- Thermal shock.

Thermal shock relates to heat transfer and as glass is a very good insulator, heat is conducted slowly across the walls when a hot liquid is filled. Another important heat related property is the dimensional change per degree change in temperature, which is low for glass. This property is also subject to the glass formulation, e.g. Pyrex is a well known type of glass with an even lower heat expansion compared with standard white flint soda glass. This is achieved by replacing some of the soda with boric oxide and increasing the proportion of silica.

It has been recognized that achieving an even distribution of glass in the walls of a container is the major factor in successfully reducing weight whilst maintaining adequate strength.

Glass Pack Design and Specification

Concept and Bottle Design

Leading glass manufacturers have state-of-the-art design expertise and systems that can be readily integrated with design house concepts to design a container which meets the requirements of branding, manufacture, filling and distribution under recommended good manufacturing practices and procedures.

The brand manager/packaging technologist can now quite readily bring together all the expertise necessary to produce a food container of the ultimate design, cost and quality to meet all their needs.

An understanding of the product specification and the filling line requirements is essential at the concept design stage. The information required includes:

- Type and density of the product.
- Carbonation level, if required, in the product.
- Closure type/neck specification required.
- Quantity to be filled.
- Type of filling process (hot-fill/cooled, hot-fill/pasteurized, ambient-fill/ sterilized or any combination of these processes).
- Is the container to be a specified volume-measuring container?
- What type of filler is to be used (volumetric or vacuum-assisted)?
- What is the filler tube size/diameter?
- Is the container to be refillable or single-trip?
- Speed of the filling operation, i.e. bottles per minute
- Impact forces on the process line.
- What pallet size is to be used in the distribution of filled stock?

- Is the depalletizer operation sweep-off or lift-off?

From this information, the glass manufacturer can select the correct finish and closure design, surface treatment requirements, the type of pack to be used for distribution to the filling line and the handling systems. Wherever possible, the body size of the container should ensure a snug fit to the pallet, since any over- hang of the glass beyond the edge of the pallet could result in breakage in transit, whilst underhang on the pallet could lead to instability. Compression, tension strapped packs can be accommodated together with live bed deliveries. This creates a highly efficient delivery system with minimal stock-holding on site, by means of just-in-time (JIT) deliveries.

Ever more challenging briefs are demanding more from packaging materials. It is well known that consumers have an innate, high quality perception of glass packaging. This emotional connection between consumer and brand is highly valued by food and drink consumers. Add to this the ability of glass to be formed into unique shapes with a wide range of decoration techniques and it is clear why glass is also the preferred choice for designers. With increased emphasis on production speed and efficiency, design freedom decreases.

Low volume: For low volume, limited or special edition products, the design freedom is high, as hand operated or semi-automated processing lines are used. Bottles may be produced using single gob machines and have a high (+0.8) capacity to weight ratio.

Main stream: For main stream production volumes, design freedom decreases, with automatic filling lines and bulk distribution being important. Bottles will be produced using larger double gob machines and have capacity to weight ratios of around 0.6–0.7.

High volume: For high volume brands, which probably have multinational distribution, the design freedom is strictly controlled to ensure compatibility with very high speed (+1000 bpm) filling lines. These brands will be produced using the largest double and triple gob NNPB machines and have capacity to weight ratios down to 0.5. A *full circle* design process creates a range of radical design options and a common sense view on the likely costs and implications of each concept. This ensures all design options are fully explored and the best design solutions are rapidly brought to the market.

Concept design: A concept design team focuses on the packaging as a brand communication tool. Using brand analysis they ensure the pack is as active as possible, at the point of sale and in use, in communicating the brand's value and positioning. Concept designers are able to work very closely with a customer's design agency, supporting the design process so that a wide range of creative options are explored, yet at the same time highlighting the practical consequences of the design options. This allows realistic, balanced decisions to be made at the earliest possible stage of the project.

Product design: Taking computer information from the concept designers, or any design agent, product designers apply a series of objective tests to the design to ensure it is fit for the purpose. These include stress analysis to check retention of carbonated products, packing line stability, and impact analysis to assess the containers' filling line performance. Strength for stacking and distribution is also checked. On completion of these tests a detailed specification of the design is issued and a 3D computer model displayed. The 3D computer model is used to create exact models for market research and to seek approval for new designs.

Mould design: The mould design team translates the product specification into mould equipment that will reproduce the container millions of times. Depending on the manufacturing plant and the process to be used, the mould equipment will vary. The level of precision required for modern glass container production is extremely exacting and has a direct effect on product quality.

The product design computer model is used to control all aspects of the design, ensuring exact replication of the design into glass. The design is now ready to be transferred to the mould makers.

Production: Quality information from each production run is fed back to the product and mould design teams to ensure best practice is used on all designs and that design teams are up to date with improvements in manufacturing capability. This closes the full circle.

Packing – Due Diligence in the use of Glass Containers

Receipt of deliveries: Glass containers are usually delivered on bulk palletized shrink-wrapped pallets. A check should be made for holes in the pallet shroud and broken glass on the pallet, and any damaged pallets rejected. The advice note should be signed accordingly, informing the supplier and returning the damaged goods.

Storage/on-site warehousing: Pallets of glass must not be stored more than six high, they must be handled with care and not shunted. Fork-lift trucks should be guarded to prevent the lift masts contacting the glass. Where air rinser cleaning is used on the filling line, the empty glass containers should not be stored outside. Pallets damaged in on-site warehousing must not be forwarded to the filling area until they have been cleared of broken glass.

Depalletization: A record should be made of the sequence and time of use of each pallet and the product batch code. Plastic shrouds must be removed with care to prevent damage to the glass; if knives are used, the blade should be shrouded at all times, so as not to damage the glass. It is necessary to ensure that the layer pads between the glass containers are removed in such a way as to prevent any debris present from dropping onto the next layer of glass. Breakages must be recorded and clean-up equipment provided to prevent any further contamination.

Cleaning Operation

Air rinse: The glass must be temperature-conditioned to prevent condensate forming on the inside, which would inhibit the removal of card- board debris. The air pressure should be monitored to ensure that debris is not suspended and allowed to settle back into the container.

On-line water rinse: Where hot-filling of the product takes place, it is essential to ensure that the temperature of the water is adequate to prevent thermal shock at the filler, i.e. not more than 60°C (140° F) differential.

Returnable wash systems: The washer feed area must be checked to ensure that the bottles enter the washer cups cleanly. A washer-full of bottles must not be left soaking overnight. In the longer term this would considerably weaken the container and could well create a reaction on the bottle surface between the hot end coating and the caustic in the washer. Where hot-filling is taking place, it is necessary to ensure that the correct temperature is reached to prevent thermal shock at the filler.

Filling operation: Clean-up instructions should be issued and displayed, so that the filling line crew know the procedure to follow should a glass container breakage occur and the need to record all breakages. It is essential to ensure that flood rinsing of the filler head in question is adequate to prevent contamination of further bottles. It is necessary to ensure that filling levels in the container comply with trading standards' requirements for measuring containers.

Capping: Clean-up instructions on the procedure to follow should breakage occur in the capper should be issued and displayed, and all breakages recorded. The application torque of the caps and vacuum levels must be checked at prescribed intervals, as must the cap security of carbonated products.

Pasteurization/sterilization: It is necessary to ensure that cooling water in the pasteurizer or sterilization retort does not exceed a differential of more than 40°C (104° F), to prevent thermal shock situations. The ideal temperature of the container after cooling is 40°C, which allows further drying of the closure and helps prevent rusting of metal closures. Air knives should be used to remove water from closures to further minimize the risk of rusting.

Labelling: Where self-adhesive labels are to be used, all traces of condensate must be eliminated to obtain the optimum conditions for label application. Adhesives must not be changed without informing the glass supplier, since this could affect the specification of adhesives/surface treatments.

Distribution: It is essential to ensure that the arrangement of the glass containers in the tray, usually plastic or corrugated fibreboard, is adequate to prevent undue movement during distribution, that the shrink-wrap is tight and that the batch coding is correct and visible.

Warehousing: The pallets of filled product must be carefully stacked to prevent isolated pockets of high loading that might create cut through in the lining compound of the container closures, as this would result in pack failures.

Quality management: The procedures of good management practice in the development, manufacture, filling, closing, processing, storage and distribution of food products in glass containers discussed in this chapter have been developed to ensure that product quality and hygiene standards are achieved along with consumer and product safety needs. Their application indicates due diligence in meeting these needs. It is essential that all procedures are clearly laid down, training is provided in their use and that regular checks are made on their implementation.

Companies can demonstrate due diligence by achieving certification under an accepted Quality Management Standard, such as ISO 9000. In the UK, the British Retail Consortium (BRC) and The Institute of Packaging (IOP) have cooperated in the publication of a Technical Standard and Protocol, which can be integrated with their ISO 9000 procedures. The BRC is a trade association representing around 90% of the retail trade and the IOP is the professional membership body, established in 1947, for the packaging industry. The IOP has amongst its objectives the education and training of people engaged in the packaging industry. This technical standard and protocol requires companies to:

- Adopt a formal Hazard Analysis System.
- Implement a documented Technical Management System.
- Define and control factory standards, product and process specifications and personnel needs.

Environmental Profile

Reuse

Glass containers can be reused for food use. However, there is only one well established household example in the UK – that of the daily doorstep delivery of fresh milk in bottles and the collection of the empty bottles. There are wide disparities in the number of trips that can be expected depending on the location, with around 12 trips per bottle being the national average. The decline of doorstep delivery has been rapid over the last decade but the system of reuse is well established. In the licensed trade, and in most places where drinks are served to customers, the drinks manufacturers operate returnable systems.

Recycling

Glass is one of the easiest materials to be recycled because it can be crushed, melted and reformed an infinite number of times with no deterioration of structure. It is the only packaging material that retains all its quality characteristics when it is recycled.

Using recycled glass, in place of virgin raw materials, to manufacture new glass containers reduces:

- The need to quarry and transport raw materials.
- The energy required to melt the glass.
- Furnace chimney emissions.
- The amount of solid waste going into landfill.

In order to recycle glass, it must first be recovered. In the UK, glass is brought by consumers to bottle banks. Currently, approximately 600 000 t/a are recovered – a figure which must increase sharply if the UK is to meet increased European Union targets for glass recovery at currently generated levels of glass container waste. Currently the recycled content of the average glass container is around 33%. The recycled proportion is higher for green than for clear containers and reflects the proportions of clear and green glass taken to bottle banks by the public. Green glass may now contain as much as 85–90% recycled glass.

Reduction – Lightweighting

In the period 1992–2002, it is claimed that the average weight of glass containers has been reduced by 40–50%. This is an average reduction. Some brand owners still retain heavy containers, e.g. spirits and liqueurs, and this causes the progress made by the glass industry to reduce the weight of packaging to be understated.

Glass Packaging Products

Plastic Packaging Materials

Plastics are made by condensation polymerization or addition polymerization (polyaddition) of monomer units. In polycondensation, the polymer chain grows by condensation reactions between molecules and is accompanied by formation of low molecular weight byproducts such as water and methanol. Polycondensation involves monomers with at least 2 functional groups such as alcohol, amine, or carboxylic groups. In polyaddition, polymer chains grow by addition reactions, in which 2 or more molecules combine to form a larger molecule without liberation of by-products. Polyaddition involves unsaturated monomers; double or triple bonds are broken to link monomer chains. There are several advantages to using plastics for food packaging. Fluid and moldable, plastics can be made into sheets, shapes, and structures, offering considerable design flexibility. Because they are chemically resistant, plastics are inexpensive and lightweight with a wide range of physical and optical properties. In fact, many plastics are heat sealable, easy to print, and can be integrated into production processes where the package is formed, filled, and sealed in the same production line. The major disadvantage of plastics is their variable permeability to light, gases, vapors, and low molecular weight molecules.

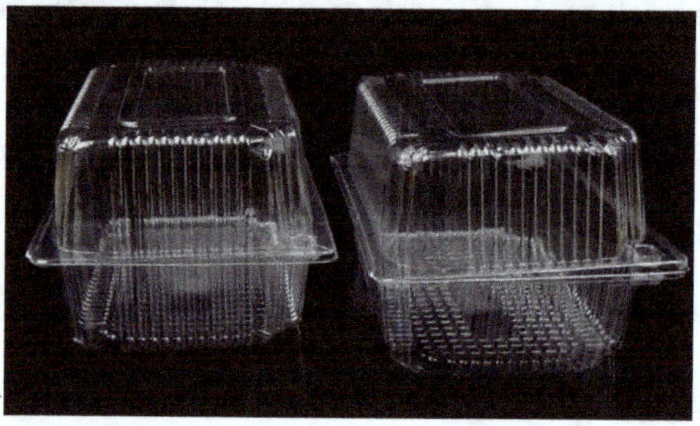

There are 2 major categories of plastics: thermosets and thermoplastics. Thermosets are polymers that solidify or set irreversibly when heated and cannot be remolded. Because they are strong and durable, they tend to be used primarily in automobiles and construction applications such as adhesives and coatings, not in food packaging applications. On the other hand, thermoplastics are polymers that soften upon exposure to heat and return to their original condition at room temperature. Because thermoplastics can easily be shaped and molded into various products such as bottles, jugs, and plastic films, they are ideal for food packaging. Moreover, virtually all thermoplastics are recyclable (melted and reused as raw materials for production of new products), although separation poses some practical limitations for certain products. The recycling process requires separation by resin type as identified by the American Plastics Council.

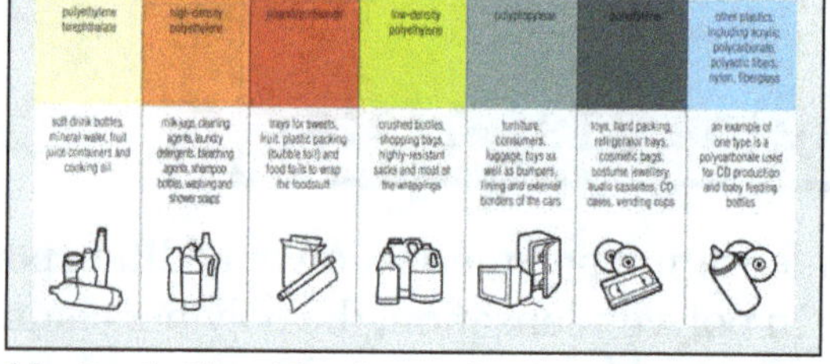

Table: Resin identification codes for plastic recycling

There have been some health concerns regarding residual monomer and components in plastics, including stabilizers, plasticizers, and condensation components such as bisphenol A. Some of these concerns are based on studies using very high intake levels; others have no scientific basis. To ensure public safety, FDA carefully reviews and regulates substances used to make plastics and other packaging materials. Any substance that can reasonably be expected to migrate into food is classified as an indirect food additive subject to FDA regulations. A threshold of regulation—defined as a specific level of dietary exposure that typically induces toxic effects and therefore poses negligible safety concerns may be used to exempt substances used in food contact materials from regulation as food additives. FDA revisits the threshold level if new scientific information raises concerns. Furthermore, FDA advises consumers to use plastics for intended purposes in accordance with the manufacturer's directions to avoid unintentional safety concerns.

Despite these safety concerns, the use of plastics in food packaging has continued to increase due to the low cost of materials and functional advantages over traditional materials such as glass and tinplate. Multiple types of plastics are being used as materials for packaging food, including polyolefin, polyester, polyvinyl chloride, polyvinylidene chloride, polystyrene, polyamide, and ethylene vinyl alcohol. Although more than 30 types of plastics have been used as packaging materials, polyolefins and polyesters are the most common.

Polyolefins: Polyolefin is a collective term for polyethylene and polypropylene, the 2 most widely used plastics in food packaging, and other less popular olefin polymers. Polyethylene and polypropylene both possess a successful combination of properties, including flexibility, strength, lightness, stability, moisture and chemical resistance, and easy processability, and are well suited for recycling and reuse.

The simplest and most inexpensive plastic made by addition polymerization of ethylene is polyethylene. There are 2 basic categories of polyethylene: high density and low density. High-density polyethylene is stiff, strong, tough, resistant to chemicals and moisture, permeable to gas, easy to process, and easy to form. It is used to make bottles for milk, juice, and water; cereal box liners; margarine tubs; and grocery, trash, and retail bags. Low-density polyethylene is flexible, strong, tough, easy to seal, and resistant to moisture. Because low-density polyethylene is relatively transparent, it is predominately used in film applications and in applications where heat sealing is necessary. Bread and frozen food bags, flexible lids, and squeezable food bottles are examples of low-density polyethylene. Polyethylene bags are sometimes reused. Of the 2 categories of polyethylene, high-density polyethylene containers, especially milk bottles, are the most recycled among plastic packages.

Harder, denser, and more transparent than polyethylene, polypropylene has good resistance to chemicals and is effective at barring water vapor. Its high melting point (160°C) makes it suitable for applications where thermal resistance is required, such as hot-filled and microwavable packaging. Popular uses include yogurt containers and margarine tubs. When used in combination with an oxygen barrier such as ethylene vinyl alcohol or polyvinylidene chloride, polypropylene provides the strength and moisture barrier for catsup and salad dressing bottles.

Polyesters: Polyethylene terephthalate (PET or PETE), polycarbonate, and polyethylene naphthalate (PEN) are polyesters, which are condensation polymers formed from ester monomers that result from the reaction between carboxylic acid and alcohol. The most commonly used polyester in food packaging is PETE.

Polyethylene terephthalate: Formed when terephthalic acid reacts with ethylene glycol, PETE provides a good barrier to gases (oxygen and carbon dioxide) and moisture. It also has good resistance to heat, mineral oils, solvents, and acids, but not to bases. Consequently, PETE is becoming the packaging material of choice for many food products, particularly beverages and mineral waters. The use of PETE to make plastic bottles for carbonated drinks is increasing steadily (van Willige and others 2002). The main reasons for its popularity are its glass-like transparency, adequate gas barrier for retention of carbonation, light weight, and shatter resistance. The 3 major packaging applications of PETE are containers (bottles, jars, and tubs), semirigid sheets for thermoforming (trays and blisters), and thin-oriented films (bags and snack food wrappers). PETE exists both as an amorphous (transparent) and a semicrystalline (opaque and white) thermoplastic material. Amorphous PETE has better ductility but less stiffness and hardness than semicrystalline PETE, which has good strength, ductility, stiffness, and hardness. Recycled PETE from soda bottles is used as fibers, insulation, and other nonfood packaging applications.

Polycarbonate: Polycarbonate is formed by polymerization of a sodium salt of bisphenol acid with carbonyl dichloride (phosgene). Clear, heat resistant, and durable, it is mainly used as a replacement for glass in items such as large returnable/refillable water bottles and sterilizable baby bottles. Care must be taken when cleaning polycarbonate because using harsh detergents such as sodium hypochlorite is not recommended because they catalyze the release of bisphenol A, a potential health hazard. An extensive literature analysis by vom Saal and Hughes (2005) suggests the need for a new risk assessment for the low-dose effects of this compound.

Polyethylene naphthalate: PEN is a condensation polymer of dimethyl naphthalene dicarboxylate and ethylene glycol. It is a relatively new member of the polyester family with excellent performance because of its high glass transition temperature. PEN's barrier properties for carbon dioxide, oxygen, and water vapor are superior to those of PETE, and PEN provides better performance at high temperatures, allowing hot refills, rewashing, and reuse. However, PEN costs 3 to 4 times more than PETE. Because PEN provides protection against transfer of flavors and odors, it is well suited for manufacturing bottles for beverages such as beer.

Polyvinyl chlorid: Polyvinyl chloride (PVC), an addition polymer of vinyl chloride, is heavy, stiff, ductile, and a medium strong, amorphous, transparent material. It has excellent resistance to chemicals (acids and bases), grease, and oil; good flow characteristics; and stable electrical properties. Although PVC is primarily used in medical and other nonfood applications, its food uses include bottles and packaging films. Because it is easily thermoformed, PVC sheets are widely used for blister packs such as those for meat products and unit dose pharmaceutical packaging.

PVC can be transformed into materials with a wide range of flexibility with the addition of plasticizers such as phthalates, adipates, citrates, and phosphates. Phthalates are mainly used in nonfood packaging applications such as cosmetics, toys, and medical devices. Safety concerns have emerged over the use of phthalates in certain products, such as toys. Because of these safety concerns, phthalates are not used in food packaging materials in the United States (HHS 2005); instead, alternative nonphthalate plasticizers such as adipates are used. For example, di-(2-ethylhexyl) adipate (DEHA) is used in the manufacture of plastic cling wraps. These alternative plasticizers also have the potential to leach into food but at lower levels than phthalates. Low levels of DEHA have shown no toxicity in animals. Finally, PVC is difficult to recycle because it is used for such a variety of products, which makes it difficult to identify and separate. In addition, incineration of PVC presents environmental problems because of its chlorine content.

Polyvinylidene chloride: Polyvinylidene chloride (PVDC) is an addition polymer of vinylidene chloride. It is heat sealable and serves as an excellent barrier to water vapor, gases, and fatty and oily products. It is used in flexible packaging as a monolayer film, a coating, or part of a co-extruded product. Major applications include packaging of poultry, cured meats, cheese, snack foods, tea, coffee, and confectionary. It is also used in hot filling, retorting, low-temperature storage, and modified atmosphere packaging. PVdC contains twice the amount of chlorine as PVC and therefore also presents problems with incineration.

Polystyrene: Polystyrene, an addition polymer of styrene, is clear, hard, and brittle with a relatively low melting point. It can be mono-extruded, co-extruded with other plastics, injection molded, or foamed to produce a range of products. Foaming produces an opaque, rigid, lightweight material with impact protection and thermal insulation properties. Typical applications include protective packaging such as egg cartons, containers, disposable plastic silverware, lids, cups, plates, bottles, and food trays. In expanded form, polystyrene is used for nonfood packaging and cushioning, and it can be recycled or incinerated.

Polyamide: Commonly known as nylon (a brand name for a range of products produced by DuPont), polyamides were originally used in textiles. Formed by a condensation reaction between diamine and diacid, polyamides are polymers in which the repeating units are held together by amide links. Different types of polyamides are characterized by a number that relates to the number of carbons in the originating monomer. For example, nylon-6 has 6 carbons and is typically used in packaging. It has mechanical and thermal properties similar to PETE, so it has similar usefulness, such as boil-in bag packaging. Nylon also offers good chemical resistance, toughness, and low gas permeability.

Ethylene vinyl alcohol: Ethylene vinyl alcohol (EVOH) is a copolymer of ethylene and vinyl alcohol. It is an excellent barrier to oil, fat, and oxygen. However, EVOH is moisture sensitive and is thus mostly used in multilayered co-extruded films in situation where it is not in direct contact with liquids.

Laminates and co-extrusions: Plastic materials can be manufactured either as a single film or as a combination of more than 1 plastic. There are 2 ways of combining plastics: lamination and co-extrusion. Lamination involves bonding together 2 or more plastics or bonding plastic to another material such as paper or aluminum (as discussed in the section on metal). Bonding is commonly achieved by use of water-, solvent-, or solids-based adhesives. After the adhesives are applied to 1 film, 2 films are passed between rollers to pressure bond them together. Lamination using laser rather than adhesives has also been used for thermoplastics. Lamination enables reverse printing, in which the printing is buried between layers and thus not subject to abrasion, and can add or enhance heat sealability.

In co-extrusion, 2 or more layers of molten plastics are combined during the film manufacture. This process is more rapid (requires 1 step in comparison to multiple steps with lamination) but requires materials that have thermal characteristics that allow co-extrusion. Because co-extrusion and lamination combine multiple materials, recycling is complicated. However, combining materials results in the additive advantage of properties from each individual material and often reduces the total amount of packaging material required. Therefore, co-extrusion and lamination can be sources of packaging reduction.

Plastic Packaging Products

Plastic Bag

A poly bag, also known as a pouch or a plastic bag, is manufactured out of flexible, thin, plastic film fabric. It is one of the common types of packaging and can carry a wide range of products including food items, flowers, waste, chemicals, magazines, and so on.

Poly bags are durable yet lightweight, reusable and flexible. Since poly bags are structurally simple to make, it can be fully customized in design, style & sizes but still remain cost-effective. Plastic recycling is also possible with poly bags, depending on the construction. Most of the poly bags are made with security features, tape attachments, hanging holes, and carrying handles to make sure the products are well secured and visually appealing to the customer.

Plastic Boxes

Plastic is used in a wide range of products, from spaceships to paper clips. A number of traditional materials, such as wood, leather, glass, ceramic, and so on, have already been replaced by plastic.

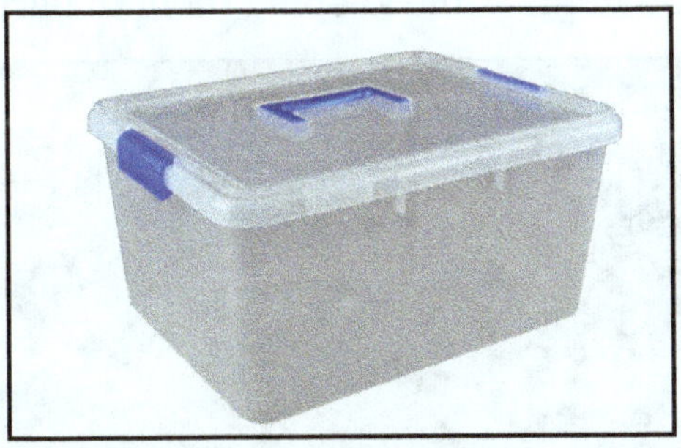

Plastic box packaging has many advantages in which they can be recycled, and generally they are much more durable than paperboard boxes. Airtight plastic packaging containers can help to preserve the quality of food and eliminate any contamination issues. Plastic packaging also does not break easily and can be stored with food under extreme conditions.

Another reason why plastic is a popular choice for packing material is because of its ability to showcase the product at any angle without necessarily opening the packaging. It is also flexible, lightweight and can be applied with films or coating to enhance packaging appearance.

Contrary to popular belief, plastic is in fact recyclable, in the sense that it takes less energy to produce new plastic, compared to glass, and other materials. Best of all, it is very cost effective.

Boil-in-bag

Boil bags or boil in bags as they are called are mainly vacuum sealed food bags that are used for preserving food items longer in a freezer. However, these bags are also extensively used in different parts of the world by many chefs for the purpose of cooking. The current models of boil bags are ideal for both cooking as well as storage. A wide range of items can be cooked in boil bags, including rice, ham, fish, salmon, beef, veal shank, zucchini, artichokes and other delicacies. They can be used for cooking when you go for weekend camping trips and they are also ideal for large scale commercial kitchen cooking. Easy to use, flexible and affordable, they can be used anywhere for the best results.

One of the best advantages of using boil bags for cooking and keeping food is that they completely eradicate the need for having additional cooking pans or storage containers. All one needs to have is hot boiling water and these bags can be used to make the food preparation that one wants. These bags are made out of industry approved materials that do not contain BPA or dioxin. They also come with special vacuum seals that prevent moisture and air from entering the bags and ruining the food. Many a times, keeping food in the freezer actually ruins the taste and flavor of the food even though it protects them from becoming stale. However, using boil bags perfectly preserve the flavors and taste of the food items while keeping them in perfect condition.

As leading restaurants as well as food and beverage stores are now acknowledging the benefits of using boil bags, more and more people are using them for their business and personal lives. Most companies manufacturing these bags offer them at wholesale prices, which mean that they are extremely cheap and affordable. Getting them in bulk also allows users to cook and store their food items in a more efficient and organized manner. Another distinct advantage that these boil bags have compared to conventional plastic storage containers or plastic bags is that they prevent the food from being exposed to oxygen and moisture. Oxygen and moisture, when combined with food can frequently give rise to the growth of mold and bacteria. On the other hand, making use of boil in bags offers a highly effective way to prevent the moisture and oxygen from damaging the food. The high level of safety standards followed by the makers of these bags has led them FDA and USDA approvals for widespread commercial usage.

The boil bags that are currently available in the market are not just suitable for home cooking; they can also be used for preparing gourmet meals in restaurants and also for catering purpose. These bags make use of sous vide cooking, which is a special type of cooking ideal for chicken and steak items while preserving their tender and juicy flavors. This enhanced flexibility with cooking is what makes professional chefs recommend boil bags for a large variety of preparations.

Edible and Biobased Packaging Materials

Biobased food packaging materials are derived from renewable sources and are potentially biodegradable that is composting (which is a technique for waste management). Biobased packaging materials include both edible coatings and edible films along with primary and secondary packaging materials. At the turn of the last century most non-fuel industrial products; dyes, inks, paint, medicines, chemicals, clothing, synthetic fibers and plastics were made from biobased resources. During the last years, the leading world research teams have been working on developing new biodegradable and edible packaging based on renewable biological sources, the so called "regulated life cycle materials". By the 1970s petroleum derived materials had, to a large extent, replaced those materials derived from natural resources. Recent developments are raising the prospects that naturally derived resources again will be a major contributor to the production of industrial products. Biobased /green polymers in food packaging are the wave of the future. The Scientific challenge is to find such applications and thus to create the demand for large scale production of biopolymers/ biomaterials that would help in attaining the sustainable development of green materials in contrast to petroleum.

References

- Excellent-benefits-of-using-aluminium-foils-for-packaging-055311: boldsky.com, Retrieved 12 April 2020
- Molded-pulp-packaging-pulp-products: advancedpaper.com, Retrieved 20 April 2020
- Biobased-Packaging-Materials-for-the-Food-Industry-258769065: researchgate.net, Retrieved 27 June 2020
- The-Environmental-Impacts-of-Packaging-229796182: researchgate.net, Retrieved 28 May 2020
- Boil-bags-indispensable-to-cooking-storing-food: universalplastic.com, Retrieved 21 June 2020
- Function-advantage-egg-trays: eggtraymachines.net, Retrieved 13 March 2020

5

Packaging of Various Types of Food

Packaging of different types of food depend on their individual characteristics and requirements. Packaging of confectionary foods should protect the food from contaminating agents, rancidity, color and aroma loss, and loss of shape. Bags, crates, baskets, cartons and hampers are suitable for packaging of fruits and vegetables. The aim of this chapter is to explore the packaging of various types of food such as beverages, food grains, dairy products, etc.

Packaging of Fruits and Vegetables

Fruit and vegetables play an important role in healthy nutrition and are high on the list of consumer priorities. However the major obstacle of purchasing ready-to-eat fresh-cut fruits and vegetables is their short shelf life, leading to quick degeneration and decomposition of the product and undesirable look and negative palatability. Fruit and vegetables are living products undergoing a ripening and at the end an ageing process, in which the plant tissue is broken down. The products undergo various biological processes, which also continue after the products have been harvested. For example there is wide variation in respiration rates and ethylene production between fruits and vegetables.

Table: Respiration rates of selected commodities.

Commodities	Respiration Rate at 5 °C (mL CO_2 kg^{-1} h^{-1})
Dried fruits and vegetables, nuts	<5
Apple, grape, potato (mature), onion, garlic	5–10
Apricot, carrot, cabbage, cherry, lettuce, peach, plum, pepper, tomato	10–20
Blackberry, cauliflower, lima bean, raspberry, strawberry	20–40
Brussels sprouts, green onion, snap bean	40–60
Asparagus, broccoli, mushroom, pea, spinach, sweet corn	>60

Conventional Packs

It is essential to minimise physical damage to fresh produce in order to obtain optimal shelf-life. The use of suitable packaging is vital in this respect. The most common form of packaging in this sector is the use of the fibreboard carton; however, for most produce, additional internal packaging, for example tissue paper wraps, trays, cups or pads, is required to reduce damage from abrasion. For very delicate fruits, smaller packs with relatively few layers of fruits are used to reduce compression damage. Moulded trays may be used which physically separate the individual piece of produce. Individual fruits may also be wrapped separately in tissue or waxed paper. This improves the physical protection and also reduces the spread of disease organisms within a pack.

Table: Classification of fruits according to their maximum ethylene production rate.

Commodities	Ethylene production rate (μL/ kg . hr at 20 °C)
Citrus fruits, strawberry, artichoke, cauliflower, cherry, asparagus, potato, grape, leafy vegetables, root vegetables	Very low: 0.01-0.1
Pineapple, blueberry, raspberry, blackberry, cranberry, aubergine, persimmon, cucumber, olive, sweet pepper, chilli pepper, pumpkin, watermelon	Low: 0.1-1.0
Banana, fig , mango, honeydew melon, mango, tomato	Moderate: 1.0-10.0
Kiwifruit (ripe), apple, avocado, plum, peach, apricot, nektarine	High: 10.0-100.0
Exotic fruit, cherimoya	Very high: >100.0

Packaging can be a major item of expense in produce marketing, so the selection of suitable containers for commercial-scale marketing requires careful consideration. Besides providing a uniform-size package to protect the produce, there are other requirements for a container:

- It should be easily transported when empty and occupy less space than when full, e.g. Plastic boxes which nest in each other when empty, collapsible cardboard boxes, fibre or paper or plastic sacks.
- It must be easy to assemble, fill and close either by hand or by use of a simple machine.
- It must provide adequate ventilation for contents during transport and storage.
- Its capacity should be suited to market demands.
- Its dimensions and design must be suited to the available transport in order to load neatly and firmly.
- It must be cost-effective in relation to the market value of the commodity for which used.
- It must be readily available, preferably from more than one supplier.

Owing to perishability, the bulk of fresh fruit and vegetables are still sold unpacked at retail or wholesale level in perforated polyethylene (PE) bags or nets. Long term transport of these materials must remain minimal to avoid brushing and consequent decay of the loosely package product. When packaging is required at the source or when an extended storage life is desired, the packaging film should have a high gas permeability and anti-fog properties. The packaging of fresh vegetables and fruit provides the largest single use of printed PE bags. The pre-packaging of potatoes, carrots, onions, parsnips, beets, radishes etc. for market sales has become an important aspect of food distribution. Because of low water vapour permeability of PE and polypropylene (PP), both films and bags are sometimes perforated to allow the product to 'breathe'. Permeation is the key to extending fresh produce shelf life.

For most commodities light is not an important influence in their post-harvest handling. However green vegetables, in the presence of sufficient light, could consume substantial amounts of CO_2 and produce O_2 through photosynthesis. Shock and vibration leads to damage to produce cells which causes an increase in respiration and may lead to enzymes being released that will cause browning reactions to begin.

Modified Atmosphere Packaging

Polymeric films have been used to package fresh products for over 35 years, with a number of benefits, including control of water loss, protection from skin abrasion and reduced contamination of the produce during handling. They also provide a barrier to the spread of decay from one unit to another. These films will also affect the movement of respiratory gases depending on the relative permeability of the film. This can lead to the development of lowered O_2 and raised CO_2 levels within the package and, as with control atmosphere (CA) storage, this can reduce the respiration of the produce and potentially extend shelf-life.

A modified atmosphere (MA) can be created within the pack in two ways. Active modification involves the pulling of a slight vacuum within the pack and then replacing the atmosphere with the desired gas mixture. Absorbers (active packaging) of CO_2, O_2 or ethylene may be included within the pack to control the concentration of these gases. In passive modification systems, the atmosphere is attained through the respiration of the commodity within the pack. The final equilibrium atmosphere will depend on the characteristics of the commodity and the packaging film. Temperature control is extremely important with MAP, as this will influence the gas permeability properties of the film as well as the respiration rate of the product. One of the main drawbacks to MAP is the potential for O_2 levels to fall too low and cause the production of undesirable off-odours caused by fermentation of the tissues.

The MA packaging technique consists of the enclosure of respiring produce in polymeric films in which the gaseous environment is actively or passively altered to slow respiration, reduce moisture loss and decay and/or extend the shelf life of the products. Many of the films used in MAP, singly do not offer all the properties required for a modified atmosphere pack. To provide packaging films with a wide range of physical properties, many of these individual films are combined through processes like lamination and co-extrusion. There are several groupings in MAP films. Polyethylene is most commonly used to provide a hermetic seal and also as a medium of control for characteristics like anti-fogging abilities, peel ability and ability to seal through a degree of contamination. The degree to which atmospheric modification takes place in packages is dependent upon several variables such as film permeability to O_2 and CO_2, product respiration and the influence of temperature on both of these processes.

Accordingly, there has been a lot of commercial interest to develop films with high gas transmission rates. High gas transmission films are obtained by modifying the film manufacturing process so that gases such as O_2, CO_2 and water vapour exit or enter the package in a controlled manner such that aerobic respiration needs are met and desirable CO_2 and moisture levels are maintained. For most plastic films, CO_2 permeability is 2-6 times greater than that of O_2, which provides a range of CO_2/O_2 permeability ratios. This allows the selection of films of suitable rations for some products.

Thus, packaging film of the correct permeability must be chosen to realise the full benefits of MAP of fresh produce. Typically, the key to successful MAP of fresh produce is to maintain an equilibrium MA (EMA) containing 2–10 % O_2/CO_2 within the package. For highly respiring produce such as mushrooms, beansprouts, leeks, peas and broccoli, traditional films like low-density polythene (PE-LD), poly(vinyl chloride) (PVC), ethylene/vinyl acetate (EVAC), oriented polypropylene (OPP) and cellulose acetate are not sufficiently permeable. Such highly respiring produce is most suitably packed in highly permeable microperforated films.

The permeability of two commercial ceramic-filled PELD films were measured and compared with those of a plain PE-LD film. The ceramic films have higher oxygen (O_2), carbon dioxide (CO_2) and ethylene (C_2H_4) permeability, higher CO_2 to O_2 permeability ratio, and higher C_2H_4 to O_2 permeability ratio. Table gives the details of film types and permeability availability. Films that have improved rates of gas transmission by virtue of their polymeric nature are usually blends of two or three different polymers, where each polymer of a blend performs a specific function such as strength, transparency and improved gas transmission to meet certain product descriptions. Furthermore, films can be laminated to achieve needed properties.

Table: Gas permeability and water transmission rate (WTR) of polymeric film available for packaging of MAP produce.

Film	Permeability (cm³/m² d atm) for 25 µm film at 25 °C			WTR (g/m²/day/atm) at 38 °C, 90% RH
	O_2	N_2	CO_2	
Ethylene- vinyl alcohol (EVAL)	3–5	–	–	16–18
PVdC–PVC copolymer (Saran)	9–15	–	20–30	–
Low-density polythene (PE-LD)	7800	2800	42.000	18
High-density polyethylene (PE-HD)	2600	650	7600	7–10

Polypropylene cast (PPcast)	3700	680	10,000	10–12
Polypropylene, oriented (OPP)	2000	400	8000	6–7
Polypropylene, oriented, PVdC coated (OPP/PVdC)	10–20	8–13	35–50	4–5
Rigid poly(vinyl chloride) PVC	150–350	60–150	450–1000	30–40
Plasticized poly(vinyl chloride) (PVC-P)	500–30.000	300–10.000	1500–46.000	15–40
Ethylene - vinyl acetate (EVAC)	12.500	4900	50,000	40–60
Polystyrene, oriented (OPS)	5000	800	18,000	100–125
Polyurethane (PUR)	800–1500	600–1200	7000–25,000	400–600
PVdC–PVC copolymer (Saran)	8–25	2–2.6	50–150	1.5–5.0
Polyamide (Nylon-6), (PA)	40	14	150–190	84–3100

Among this class are high (6–18 %) EVAC content, PE-LD, oriented polypropylene laminates, styrene butadiene block copolymer films and ultra-low-density ethylene octene copolymer films and polyolefin plastomer octene copolymer films. A lot of work has been reported on MAP using different types of films. Commercially available MAP systems for small and large quantities of produce are given in table. Films using micro perforations can attain very high rates of gas transmission. The diameter of micro perforation generally ranges from 40 to 200 μm and by altering the size and thickness of micro perforations; gas permeability through a package can be altered to meet well defined product requirements. Based on the rates of respiration and gas transmission through the micro perforations and the base film, packages have been developed that maintain desired levels of O_2 and moisture for high respiring mushrooms.

Micro perforated films have also been used to extend the storability of strawberries and nectarines, leeks, asparagus, parsnips, cherry tomatoes, sweet corn and apple. A procedure permitting the simple determination of O_2 and CO_2 permeance of micro perforated films used to pack respiring foods under real conditions has been evaluated. Macro perforated films has also been used to improve the keeping quality of strawberries and raspberries by combining high O_2 atmospheres with low oxygen MAP. However, micro perforated films are relatively expensive, permit moisture and odour loss, and may allow for the ingress of microorganisms into sealed packs during wet handling situations.

A very interesting development for the packing of fresh prepared produce involves the use of high O_2 (70–100 %) MAP, which has been recently shown to overcome the many disadvantages of current air packing and low O_2 MAP. High O_2 MAP has been demonstrated to inhibit enzymic discolorations, prevent anaerobic fermentation reactions and inhibit microbial growth with the result of extending prepared produce shelf-life.

Different commodities have different amounts of internal air space (potatoes 1–2 %, tomatoes 15–20 %, apples 25–30 %). A limited amount of air space leads to increase in resistance to gas diffusion. One of the primary effects of MAP is a lower rate of respiration, which reduces the rate of substrate depletion. Ethylene (C_2H_4) is a natural plant hormone and plays a central role in the initiation of ripening, and is physiologically active in trace amounts (0.1 ppm). C_2H_4 production is reduced by about half at O_2 levels of around 2.5 %. This low O_2 retards produce ripening by inhibiting both the production and action of C_2H_4.

Table: Recommended gas mixtures for MAP.

Fruits	O_2 (%)	CO_2 (%)	N_2 (%)	Vegetables	O_2 (%)	CO_2 (%)	N_2 (%)
Apple	1–2	1–3	95–98	Artichoke	2–3	2–3	94–96
Apricot	2–3	2–3	94–96	Beans, snap	2–3	5–10	87–93
Avocado	2–5	3–10	85–95	Broccoli	1–2	5–10	88–94
Banana	2–5	2–5	90–96	Brussels sprouts	1–2	5–7	91–94
Grape	2–5	1–3	92–97	Cabbage	2–3	3–6	81–95
Grapefruit	3–10	5–10	80–92	Carrot	5	3–4	91–95
Kiwifruit	1–2	3–5	93–96	Cauliflower	2–5	2–5	90–96
Lemon	5–10	0–10	80–95	Chili peppers	3	5	92
Mango	3–7	5–8	85–92	Corn, sweet	2–4	10–20	76–88
Orange	5–10	0–5	85–95	Cucumber	3–5	0	95–97
Papaya	2–5	5–8	87–93	Lettuce (leaf)	1–3	0	97–99
Peach	1–2	3–5	93–96	Mushrooms	3–21	5–15	65–92
Pear	2–3	0–1	96–98	Spinach	Air	10–20	–
Pineapple	2–5	5–10	85–93	Tomatoes	3–5	0	95–97
Strawberry	5–10	15–20	70–80	Onion	1–2	0	98–99

Also metabolic processes such as respiration and ripening rates are sensitive to temperature. Biological reactions generally increase two to three-fold for every 10 °C rise in temperature. Therefore temperature control is vitally important in order for a MAP system to work effectively. Film permeability also increases as temperature increases, with CO_2 permeability responding more than O_2 permeability. Low RH can increase transpiration damage and lead to desiccation, increased respiration, and ultimately an unmarketable product. One serious problem associated with high in-package humidity is condensation on the film that is driven by temperature fluctuations. A mathematical model was developed for estimating the changes in the atmosphere and humidity within perforated packages of fresh produce. The model was based on the mass balances of O_2, CO_2, N_2 and H_2O vapours in the package. Also a procedure to maintain desired levels of O_2 and CO_2 inside packages that are exposed to different surrounding temperatures was designed and tested.

The concentrations of O_2 and CO_2 within a package can be modeled. Useful models have been developed that would allow fresh produce processors to choose packaging materials most suited to the enclosed product. A common mathematical model involves the use of what is known as a Michaelis–Menten type respiratory model to describe the influence of temperature, O_2 (and potentially CO_2) on respiration. This approach has been used for blueberries, strawberries, raspberries and apple slices. Respiratory models are then coupled with an equation describing the temperature sensitivity of film permeability to gases (known as the Arrhenius equation) to predict package O_2 partial pressure as a function of temperature, product mass, surface area and film thickness. Using polymeric films, MA packaging systems for products with low to medium respiration rates have to some extent been successfully developed. Products such as broccoli, mushrooms, leeks, etc. exhibit very high rates of respiration such that conventional films can potentially over modify the pack atmosphere and thus result in fermentation.

Qualitative changes in the broccoli (Brassica oleracea italica) under MAP in perforated polymeric film had also been studied. It has been found that perforated PP film packages (2 holes, each of 0.3 mm in diameter) and having a film area of 0.1 m^2 could be used to store broccoli for 4 days under MAP with maintenance of chlorophyll and ascorbic acid. Comparative evaluation of the effect of storage temperature fluctuation on MAP of selected fruits and vegetables like mushrooms (Agaricus bisporus), broccoli (Brassica oleracea L. cv. Acadi) and mature-green tomatoes (Lycopersicon esculentum cv. Trust) had also been done by a researcher.

The quality of the products stored under the temperature fluctuating regime was severely affected as indicated by extensive browning, loss of firmness, weight loss increase, the level of ethanol in the plant tissue, and infection due to physiological damage and excessive condensation, compared to products stored at constant temperature. It was clear that temperature fluctuation, even if it should occur only once, can seriously compromise the benefits of MAP and safety of the packaged produce. The effect of storage conditions and physical tissue damage on membrane peroxidation in minimally processed cucumber tissue was investigated.

Tissue hardness of MAP sealed samples increased for the first 3 days for all tissues but chilled tissues started to soften on the sixth day. Cucumbers packaged in perforated 31.75 μm PELD bags were found to have less severe chilling injury than non-wrapped fruit in storage at 5 °C and 90–95 % RH. The influence of MAP on the sensory characteristics and shelf life of shiitake mushrooms (Lentinula edodes) was also studied using PE-LD, PP and macro perforated film. Sensory analysis showed that mushrooms stored under MA (active and passive) had a higher deterioration rate than those stored in PP macro perforated films, and lower sensory quality values during the entire storage time. These results suggest that mushroom deterioration was probably due to shiitake mushrooms' sensitivity to high CO_2 concentrations.

Mushrooms have been packed under micro porous and hydrophilic films and stored at 10 °C and 20 °C under high relative humidity (>92 %) to study the unique gas barrier properties of hydrophilic films (wheat gluten based material and synthetic polymer). Unique steady state atmospheres, poor in both oxygen and carbon dioxide, were observed, regardless of the temperature and the hydrophilic film used, owing to their high selectivity to gas diffusion. The effect of super-atmospheric O_2 and MAP on plant metabolism, organoleptic quality and microbial growth of minimally processed baby spinach was also studied. Packaging film O_2 transmission rate and initial levels of super-atmospheric O_2 in the packages significantly affected the changes of in-package atmospheres during storage and consequently quality of baby spinach leaves. Microbial proliferation and sensory quality aspects of sliced onions were tested at different temperatures (-2, 4 and 10 °C) and atmospheric conditions (with or without 40 % CO_2/59 % N_2/1 % O_2). The microbial shelf lives of the tested onions in 40 % CO_2/59 % N_2/1 % O_2, or at -2, 4 and 10 °C, were 12.5, 9.5, 7, 12, 9 and 6 days, respectively, and their sensory shelf lives were 12, 8, 5, 10.5, 7 and 5 days, respectively.

Work on other vegetables has also been reported like sweet corn and snow pea pods. The shelf life of shelled peas packed in PE-LD (25 μm thickness) was 45, 17, 7 and 4 days when stored at the temperatures of -11, 5, 15 °C and room temperature respectively, considering the quality indices like total soluble solids, total water soluble sugars, protein, physiological weight loss and decay. The shelf life of shelled peas packaged in PE-HD was found as 4, 7, 17 and 45 days when stored at the temperatures of ambient, 15, 5 and -11 °C, respectively. The influence of different storage conditions on the storability of packaged shredded cabbage had also been studied. A higher variation in CO_2 and O_2 concentrations, and consequent accumulation of anaerobic metabolites had a negative influence on the sensorial properties of the cut cabbage.

Nitrogen (N_2) gas packaging for fresh-cut vegetables (lettuce and cabbage) has been examined as a means of MAP for extending the shelf life of cut vegetables. Degradation of cut vegetables in terms of appearance was delayed by N_2 gas packaging. Because of this effect, the appearance of fresh-cut vegetables packaged with N_2 gas remained acceptable at temperatures below 5 °C after 5 days.

The MAP combined with ozone and edible coating films were used for improving the effect of preservation of strawberry. The optimum gas composition of MAP test for strawberry was 2.5 % O_2/16 % CO_2. Also the integrated model approach was used to study the effect of MA conditions on the keeping quality of 'Elsanta' strawberries as limited by spoilage. It was observed that the quality of SO_2-free 'Superior seedless' table grapes was preserved in MAP. The improvement of the overall quality of table grapes stored under MAP in combination with natural antimicrobial compounds had also been studied. The effect of MA on the quality of many fresh-cut products has been studied. Successful applications include mushroom, apples, tomato, pineapple, butter-head lettuce, potato, kiwifruit, salad savoy, honeydew, mangoes and carrot.

Fresh-cut cantaloupe cubes were placed in film sealed containers in which the internal gas mixture was attained naturally (nMAP), was flushed with 4 kPa O_2/10 kPa CO_2 (fMAP), or was maintained near atmospheric levels by perforating the film (PFP). While both nMAP and fMAP maintained the sale able quality of melon cubes at 5 °C, fMAP maintained quality better than MAP. Also quality of fresh-cut tomato slices was compared during cold storage under various MAP conditions. MAP provided good quality tomato slices with a shelf life of 2 weeks or more at 5 °C. Effects of packaging materials and methods on the shelf life of some fruits and vegetables are given in table.

Table: Effect of packaging materials and methods on the shelf life of fruits and vegetables.

Food/Treatment	Packaging materials/methods	Shelf life
Peach, Cauliflower, Truffle	Tray: PP; Cover: PE-LD/PET(40 µm), 0-14 microperforated package, all wrapped in PE	4 days at 4 °C
Strawberry	stretch PVC	8 days at 1 °C
Minimally processed fruits (kiwi, banana and prickly pear)	1) PE/Al/PET	4-12 days at 5 °C
	2) Coex.polyolefinic high permeable film	
Sweet cherry	5 % O_2 + 10 % CO_2	80 days at 1 °C
	PE: 13-18 % O_2 + 2-4 % CO_2	40 days at 1 °C
	70 % O_2 + 0 % CO_2	20 days at 1 °C
	Air	30 days at 1°C
Cactus pear fruits	Cryovac MY 15 Plastic box	9 days at 4 °C
Carrots, minimally processed	PP+cPP/OPP in:	
	5 % O_2 / 10% CO_2 / 85 % N_2	2 days at 4 °C
	80 % O_2 / 10 % CO_2 / 10 % N_2	7 days at 4 °C

Cabbage, shredded	OPP (30 µm)	9-10 days at 3 °C
Cabbage, shredded	Glass jar; PE (30 µm); PP (30 µm) in:	7 days at 0 and 10 °C
	Air	
	100 % N_2,	
	MAP 1: 100 % N_2,	
	MAP 2: 5 % O_2/95 % N_2,	
	MAP 3:10 % O_2/90 %	
	MAP 4: 70 % O_2/30 % N_2 and 100 % O_2	
Lettuce (shredded)	OPP (30 µm)	8 days at 4°C
Spinach Fresh-cut	PE-LD	8 days at 4 °C and 90 % HR
	OPP	6 days at 4 °C and 90 % HR
Lettuce	PE	10 days in the dark at 7 °C.a
Broccoli	PE-non-perforated	13 days at 4 °C 3 days at 20 °C
	PE-2 holes (750 µm in dia.)	23 days at 4 °C
	PE-4 holes (8.8 mm in dia.)	5 days at 20 °C
	Unwrapped broccoli	10 days at 4 °C, 2.5 days at 20 °C
Fresh-cut vegetables, fennels cut lettuce salads and shredded carrots	PA/PE (102 µm)	
	Air	less than 5 days at 4 °C
	Vacuum	13.5 at 4 °C
	MAP: 70 % N_2/30 % CO_2	5.8 at 4 °C
Cherry tomatoes ozone treatment (5–30 mg/l ozone gas for 0–20 min)	CA 1: 5 % CO_2/17 % O_2 (7°C)	20 days at 7 °C
	CA 2: 5 % CO_2/15 % O_2 (22°C)	10 days at 22 °C
	PE-LD (50 µm); MAP	
	Glass jars	
Melon, minimally fresh processed	MAP passive in:	10 days at 5 °C
	PP mikroperforated	
	BOPP	
	OPP	
	PP-makroperforated (control)	

Blueberry nectar Aseptically Filled (93–97 °C for 30 s)	PET 1: Standard	9-12 months at 4-20 °C
	PET 2: O_2-absorber	
	PET 3: 5-layers	
	PET 4: 3-layers	
Tomato	PE-LD (50 mm) MAP + O_2-absorber	8 days at 20 °C

Edible Coatings and Films

The application of edible coatings appears to be one of the most innovative approaches to extend the commercial shelf life of fruits and vegetables by, among other mechanisms, acting as a barrier against gas transport and showing similar effects to storage under controlled atmospheres. Indeed, over the last two decades the development and use of bio-based packaging materials to prolong the shelf-life and improve the quality of fresh products has been receiving increased attention. The reasons for such an interest are mainly related to environmental issues due to disposal of conventional synthetic food-packaging materials. However, in order such edible films and coatings to be used at a commercial level in food products they must fulfill some basic requirements: acceptable sensorial characteristics, appropriate barrier properties, good mechanical strength, reasonable microbial, biochemical and physicochemical stability, safety, low cost and simple technology for their production. The effectiveness of edible coatings for protection of fruits and vegetables also depends on controlling the wettability of the coating solutions, which affects the coating thickness. Thus, edible coating formulations must wet and spread uniformly on the vegetable's surface and, after drying, a coating that has adequate adhesion, cohesion and durability to function properly must be formed.

Table: Oxygen and carbon dioxide permeabilities of edible films.

Film	Thickness (mm)	Permeability at 0% RH (10^{-15} l/m² s Pa)		Permeability ratio (CO_2/O_2)
		O_2	CO_2	
Corn-zein	0.12-0.31	0.36 30 °C	2.67 21 °C	7.5
Wheat gluten	0.23-0.42	0.20 30 °C	2.13 21 °C	9.5
Methyl cellulose Low level (MC (L))	0.04-0.07	2.17 30 °C	69.00 21 °C	31.6
Hydroxypropylcellulose Low level (HPC (L))	0.05	3.57 30 °C	143.99 21 °C	40.6
HPC/Lipids	0.15	3.44 30 °C	81.75 21 °C	23.7

Cozeen	0.09	0.89 37,8 °C	5.25 22,8 °C	5.9
Wheat gluten	0.14	0.09 37,8 °C	0.03 22,8 °C	0.3
Corn-zein	0.08	0.16 25 °C	-	-
Wheat gluten	0.15	0.08 25 °C	-	-

Edible films and coatings are generally based on biological materials such as proteins, lipids and polysaccharides. The main polysaccharides that can be included in edible coating formulations are starch and starch derivatives, cellulose derivatives, chitosan, pectin, alginate and other gums. Carboxymethyl-cellulose is a cellulose derivative that has received considerable attention with several examples of applications in many fruits and vegetables. Some polysaccharides have been used as edible coatings to improve the quality of different fresh-cut fruit.

A commercial edible coating formulation based on carboxymethylcellulose and sucrose fatty acid esters, has been applied to pears, cherries and asparagus. Pullulan, an extracellular polysaccharide produced by Aureobasidium pullullans, also is capable of forming edible films but has not been largely exploited as a coating material in fruits and vegetables, presumably because of its high water solubility. One example of pullulan used as a coating hydrocolloid was for strawberries and kiwifruit. Proteins that can also be used in formulations of edible coatings for fruits and vegetables include those derived from animal sources, such as casein and whey proteins, or obtained from plant sources, like corn zein, wheat gluten and soy protein. Whey protein based coatings have been extensively used to extend the shelf life of fruits and vegetables.

Antimicrobial Packaging

The use of edible coatings as carriers of antimicrobial compounds is another potential alternative to enhance the safety of fresh-cut produce. Antimicrobial edible coatings may provide increased inhibitory effects against spoilage and pathogenic bacteria by maintaining effective concentrations of the active compounds on the food surfaces. There are several categories of antimicrobials that can be potentially incorporated into edible coatings, including organic acids (acetic, benzoic, lactic, propionic, sorbic), fatty acid esters (glyceryl monolaurate), polypeptides (lysozyme, peroxidase, lactoferrin, nisin), plant essential oils (cinnamon, oregano, lemongrass), nitrites and sulphites, among others. Although several types of antimicrobials incorporated into edible coatings have been used for extending shelf-life of fresh commodities, their use in fresh-cut fruits is yet limited. Currently, organic acids and plant essential oils are the main antimicrobial agents incorporated into edible coatings for fresh-cut fruits. Despite the good results achieved so far with the incorporation of essential oils into edible coatings, the major drawback is their strong flavour which could change the original taste of foods.

To confer antimicrobial activity, antimicrobial agents may be coated, incorporated, immobilised, or surface modified onto package materials. The classes of antimicrobials listed range from acid anhydride, alcohol, bacteriocins, chelators, enzymes, organic acids and polysaccharides. Examples of commercial antimicrobial materials in the form of concentrates (e.g. AgIONe, AgION Technologies LLC, USA) extracts (Nisaplin (Nisin), Integrated Ingredients, USA) and films (Microgarde Rhone-Poulenc, USA) were also presented. Antimicrobial packages have had relatively few commercial successes except in Japan where Ag-substituted zeolite is the most common antimicrobial agent incorporated into plastics. Ag ions inhibit a range of metabolic enzymes and have strong antimicrobial activity. Antimicrobial films can be classified into two types: those that contain an antimicrobial agent which migrates to the surface of the food and, those which are effective against surface growth of microorganisms without migration. Coating of films with antimicrobial agents can result in effective antimicrobial activity. Researchers carried out a study to evaluate the potential use of packaging materials as delivery vehicles for carrying and transferring nisin-containing formulations onto the surfaces of fresh poultry products.

The antimicrobial activity of the fabricated multilayer films was also evaluated using an agar plate diffusion method. It was reported that coating the PE film with grapefruit seed extract (GFSE) with the aid of a polyamide binder resulted in greater antimicrobial activity compared to GFSE incorporation by co-extrusion. Using the agar diffusion test, the co-extruded film with 1% w/w GFSE showed antimicrobial activity against Metaphycus flavus only, whereas a film coated with 1% GFSE showed activity against several microorganisms such as Escherichia coli, Staphylococcus aureus and Bacillus subtilis. There is a growing interest in edible coatings due to factors such as environmental concerns, the need for new storage techniques and opportunities for creating new markets for underutilised agricultural commodities with film forming properties. Edible coatings and films prepared from polysaccharides, proteins and lipids have a variety of advantages such as biodegradability, edibility, biocompatibility, aesthetic appearance and barrier properties against oxygen and physical stress.

Antimicrobial packaging can play an important role in reducing the risk of pathogen contamination, as well as extending the shelf-life of foods; it should never substitute for good quality raw materials, properly processed foods and good manufacturing practices. It should be considered as a hurdle technology that in addition with other non-thermal processes such as pulsed light, high pressure and irradiation could reduce the risk of pathogen contamination and extend the shelf-life of perishable food products. Participation and collaboration of research institutions, industry and government regulatory agencies will be a key on the success of antimicrobial packaging technologies for food applications.

Active Packaging

Active packaging is an emerging and exciting area of food technology that is developing owing to advances in packaging technology, material science, biotechnology and new consumer demands. This technology can confer many preservation benefits on a wide range of ambient-stable and chilled food products. The intention is to extend the shelf-life of foods, whilst at the same time maintaining nutritional quality and assuring microbial safety. The use of active packaging is becoming increasingly popular and many new opportunities will open up for utilising this technology in the future.

Active packaging employs a packaging material that interacts with the internal gas environment to extend the shelf-life of a food. Such new technologies continuously modify the gas environment (and may interact with the surface of the food) by removing gases from or adding gases to the headspace inside a package.

1. Ethylene scavenging: A chemical reagent, incorporated into the packaging film, traps the ethylene produced by ripening fruit or vegetables. The reaction is irreversible and only small quantities of the scavenger are required to remove ethylene at the concentrations at which it is produced.

2. Oxygen scavenging: The presence of oxygen in food packages accelerates the spoilage of many foods. Oxygen can cause off-flavour development, colour change, nutrient loss and microbial attack. One of the most promising applications of oxygen scavenging systems in food packages is to control mould growth.

Table: Properties of major types of Ageless oxygen scavengers.

Type	Function	a_w	Absorption speed	Capacities
Z	Decreases O_2	< 0.65	1 – 4 days	20 – 2 000 cc
S	Decreases O_2	> 0.65	0.5 – 2 days	500 – 2000 cc
FX	Decreases O_2	> 0.85	0.5 – 1 days	20 – 300 cc
E	Decreases O_2 Decreases CO_2	> 0.30	3 – 8 days	O_2: 25–200 cc CO_2: 250–2000 cc
G	Decreases O_2 Increases CO_2	0.3 - 0.5	1 – 4 days	NA

3. Humidity control: Condensation or 'sweating' is a problem in many kinds of packaged fruit and vegetables. When one part of the package becomes cooler than water is likely to condense in the cooler areas.

4. Carbon dioxide release: High carbon dioxide levels are desirable in some food packages because they inhibit surface growth of microorganisms. Fresh meat, poultry, fish, cheeses and strawberries are foods which can benefit from packaging in a high carbon dioxide atmosphere. However with the introduction of modified atmosphere packaging

there is a need to generate varying concentrations of carbon dioxide to suit specific food requirements. Since carbon dioxide is more permeable through plastic films than is oxygen, carbon dioxide will need to be actively produced in some applications to maintain the desired atmosphere in the package.

5. Sulphur dioxide: Serious loss of table grapes can occur unless precautions are taken against mould growth. It is necessary to refrigerate grapes in combination with fumigation using low levels of sulphur dioxide. Fumigation can be conducted in the fruit cool stores as well as in the cartons. Carton fumigation consists of a combination of quick release and slow release systems which emit small amounts of sulphur dioxide. When the temperature of the carton grapes rises due to inadequate temperature control, the slow release system fails releasing its entire sulphur dioxide quickly. This can lead to illegal residues in the grapes and unsightly bleaching of the fruit.

Flexible packaging materials such as PE-LD and linear low density polyethylene (PE-LLD) when impregnated with potassium permanganate and cinnamic acid, respectively become ethylene scavengers. Fresh fruits and vegetables, such as mango, tomato, banana, and papaya exhibit more shelf-life of two to three weeks when packed in such ethylene scavenging films. The incorporation of ethylene scavengers improves the physico-mechanical properties of the packaging materials considerably.

Design of active modified atmosphere packaging for vegetables was studied by developing a mathematical model predicting gas changes and based on the following independently evaluated parameters: vegetable respiration rate, film permeability, and oxygen absorption kinetic of the absorber. A step-by-step model validation was performed on a tomatoes/low-density polyethylene pouch/commercial iron-based scavenger system at 20 °C.

Smart or Intelligent Packaging

Smart packaging uses features of high added value that enhance the functionality of the product, notably mechanical, electronic and responsive ink features, for example electronic and mechanical dispensers in which drugs are supplied and the prepared meal that automatically tells the microwave how it should be cooked. Smart packaging can be categorised into two types, those which incorporate integrated circuits (IC's) and does which do not incorporate IC's otherwise known as chipless smart packaging. Packaging that incorporate diagnostic indicators are also included in smart packaging. These can be used for such functions as monitoring vibration, acidity, tilt, shock, humidity, light, heat, time chemicals, virus or bacteria as they develop or as they are contacted.

Fresh-cut produce continues to be one of the fastest growing segments of food retailing and while conventional film packaging is suitable for lettuce and prepared salads, it cannot cope with the high respiration rates of pre-cut vegetables and fruit, leading to early product deterioration. Novel breatheable polymer films are already in commercial use for fresh-cut vegetables and fruit. Packaging films that are acrylic side-chain crystallisable polymers tailored to change phase reversibly at various temperatures from 0-68 °C are available. As the side chain components melt, gas permeation increases dramatically, and by further tailoring the package and materials of construction, it is possible to fine tune the carbon dioxide to oxygen permeation ratios for particular products. The final package is 'smart' because it automatically regulates oxygen ingress and carbon dioxide egress by transpiration according to the prevailing temperature. In this way, an optimum atmosphere is maintained around the product during storage and distribution, extending freshness and allowing shipping of higher quality products to the consumer.

Intelligent packaging can change colour to let the customer know how fresh the food is and show if the food has been spoiled because of a change in temperature during storage or a leak in the packaging. Time temperature integrators (TTI's) are devices that show an irreversible change in a physical characteristic, usually color or shape, in response to temperature history. The TTI's are expected to mimic the change of a certain quality parameter of the food product undergoing the same exposure to temperature. The TTI's presently on the market have working mechanisms based on different principles: biological, chemical and physical. For the first type, the change in biological activity, such as microorganisms, spores or enzymes is the basic working principle. The others are based on a purely chemical or physical response towards time and temperature, such as an acid-base reaction, melting, polymerization, etc.

Fresh-Check LifeLines integrator is supplied as self-adhesive labels, which may be applied to packages of perishable products to assure consumers at point-of-purchase and at home that the product is still fresh. It is commonly referred to as having a bull's eye configuration. It is a full history indicator whose working mechanism is based on the color change of a polymer formulated from diacetylene monomers. It consists of a small circle of polymer surrounded by a printed ring for color reference. The polymer, which starts lightly colored, gradually deepens in color to reflect the cumulative exposure to temperature. The polymer changes color at a rate proportional to the rate of food quality loss: the higher the temperature, the more rapidly the polymer changes in color.

Vitsab Indicator is a full history integrator based on an enzymatic reaction. The device consists of a bubble-like dot containing two compartments: one for the enzyme solution, lipase plus a pH indicating dye compound and the other for the substrate, consisting primarily of triglycerides. The dot is activated at the beginning of the monitoring period by application of pressure on the plastic bubble, which breaks the seal between compartments. The ingredients are mixed and as the reaction proceeds a pH change results in a color change. The dot, initially green in color, becomes progressively yellow as product approaches the end of shelf-life. The reaction is irreversible and will proceed faster as temperature is increased and slower as temperature is reduced. The integrator, in the format of adhesive labels, is available in two basic configurations: single or triple-dot. Single-dot tags are used for transit temperature monitoring of cartons and pallets of product and for consumer packages as well. ripeSense™ is the world's first intelligent ripeness indicator label. ripeSense™ evolved from the simple idea of making a fruit label that is capable of more than just branding product and this has led to the next revolution in fresh produce marketing.

The 3M Monitor Mark uses a coloured ester and phthalate mix with the desired melting point that is coloured with a blue dye. Above its melting point it diffuses along a wick, and the progress along this wick gives an indication of how long the indicator has been liquid. A polyester strip keeps the liquid away from the wick, until it is pulled by the operator to start the device. It is believed that tomorrows food packages will certainly include radio frequency identification (RFID) tags. RFID tags are an advanced form of data information carrier that can identify and trace a product.

Biodegradable Packaging

Currently, there are several types of bio-based polymers on the market: same coming from petrochemical monomer, like certain types of polyester, polyester amides and polyvinyl alcohol, produced by different manufacturer, used principally as films or moulding. Four other bio-based polymers are starch materials, cellulose materials, polylactic acid (Polyester, PLA), polyhydroxy acid (polyester, PHA). Until now, the PHA polymer is a very expensive polymer because it is commercially available in very limited quantities. PLA is becoming a growing alternative as a green food packaging materials because it was found that in many situations it performs better than synthetic ones, like oriented polystyrene (OPS) and PET materials.

There is an increasing demand for identifying biodegradable packaging materials and finding innovative methods to make plastic degradable. Biodegradation is the process by which carbon-containing chemical compounds are decomposed in the presence of enzymes secreted by living organisms. The use of bioplastic is to replicate the life cycle of biomass by conserving the fossil fuels, carbon dioxide and water production. There are three requirements for the fast degradation process viz. temperature, humidity and type of microbes.

In the short term, biobased materials will most likely be applied to foods requiring short-term chill storage, such as fruits and vegetables, since biobased materials present opportunities for producing films with variable CO_2/O_2 selectivity and moisture permeability. However, to succeed, biobased packaging of foods must be in compliance with the quality and safety requirements of the food product and meet legal standards. Additionally, the biobased materials should preferably preserve the quality of the product better and longer to justify any extra material cost.

Application of Nanocomposites

Nanocomposite materials are composed of nanoscale structure that enhances the macroscopic properties of food products. The common nanocomposites used in the food packaging industry are (i) Polymer clay nanoclay (ii) Silica nanocomposites of nanosilver. The effects of nanoclay in polymers are increased stiffness, strength, nucleating agent in foams, smaller cell size, higher cell density, and flame retardant. Nanosilver is composed of de-ionized water with silver in suspension which has excellent antibacterial properties. Silver nanoparticles interact well with other particles as they have large surface area relative to volume which increases their antibacterial efficiency as a result of which they are extensively used in the food packaging industry.

Table: Properties of some biodegradable plastics.

Material	Film preparation	Moisture barrier	Oxygen barrier	Mechanical properties
Starch/Polyvinyl alcohol (PVAL)	Extrusion	-	+	+
Polyhydroxybutyrate/valerate (PHB/V)	Extrusion	+	+	+/-
Polylactic acid (PLA)	Extrusion	+/-	-	+

Although application of nanotechnology in the food industry started later than other industries, the great potential of food nanotechnology, especially in the area of improving food quality and securing food safety, has been recognised by many nanoscientists and technologists. Nanotechnology is already applied to the food and food packaging industries.

One of the potential applications of nanotechnology in food packaging is polymer/clay nanocomposites; they have recently emerged due to their potential for improving properties of packaging materials such as increased mechanical, barrier and chemical properties with a small amount (less than 5% by weight) of nanoclays reinforcement. However, most work done on polymer/clay nanocomposites has focused mainly on synthetic polymers.

The functional properties of biopolymer-based edible films or coatings have been shown to act as barrier to solute and gas and enhance food quality and shelf life. However, these films do not display good mechanical and water vapor barrier properties due to their hydrophilic characteristics. To overcome these issues, a new approach has been developed, which use hybrid materials consisting of polymers and layered silicates. Layered silicates, such as montmorillonite (MMT) clay mineral, result from the stacked arrangement of negatively charged silicate layers and contain a platelet thickness of about 1 nm with a high aspect ratio (ratio of length to thickness). The layered silicate filled polymer composites exhibit extraordinary enhancement of mechanical, thermal and physicochemical properties at a low level of filler concentration in comparison to pure polymer and conventional microcomposites. In particular, these nanocomposites have excellent barrier properties because the presence of clay layers delays the diffusing molecule pathway due to tortuosity. Some of the works done with biopolymer-based nanocomposites were based on starch or polysaccharides, such as wheat and maize starch, thermoplastic starch, and chitosan. A few studies on protein-based nanocomposites including soy protein, whey protein, and wheat gluten. Most of the biopolymer-based nanocomposites have shown appreciable improvements in mechanical and barrier properties compared to the counterpart biopolymer films.

Whey protein has received much attention for its potential use as an edible film and coating because it has been shown to make transparent films and coatings that can act as excellent oxygen barriers and provide certain mechanical properties. Unlike chitosan film, whey protein films have not shown any antimicrobial activity; therefore, incorporation of antimicrobial agents, such as sorbic acid, p-aminobenzoic acid, and lysozyme, is needed to impart this property. Recently, researcher found that chitosan based nanocomposite films blended with some organically modified MMT, such as Cloisite 30B, exhibited antimicrobial activity against Gram-positive bacteria. They postulated that the antimicrobial action came from the quaternary ammonium salt of the organically modified nano-clay.

Packaging of Meat and Poultry

The meat industry is an important sector of food industry in the world and comes in the first five ranked agricultural commodities like rice, milk, wheat etc.

Importance of Meat Packaging

Food packaging is an integral part of food processing and a vital link between the processor and the eventual consumer for the safe delivery of the product through the various stages of processing, storage, transport, distribution and marketing. All over the world, the consumers are showing greater awareness towards food packaging as it provides awareness on the quality, quantity and the hygienic standards. A very important aspect of meat preservation is the suitable packaging of the product. The main purpose of packaging is to protect meat/meat product from microbial contamination, effect of light, oxygen or any physical damage or chemical changes. The selection of the packaging material has to be done very carefully to protect the different physico-chemical properties like nature of pigments, sensory attributes and microflora.

The purpose is to retard or prevent the main deteriorative changes and make the products available to the consumers in the most attractive form. However, initial quality of the meat has to be very good because packaging can only maintain the existing quality of meat or delay the onset of spoilage by controlling the factors that contribute to it. The product, therefore, is only protected for a limited period determined by the system that is used. Thus, meat and meat products need a specialised package profile depending upon the type of processing, condition of storage and distribution.

Packaging of Fresh Meat

Packaging Requirements

Fresh meat is highly perishable and a biologically active item. The quality of fresh meat is affected by the growth of micro-organisms, enzyme activity and by oxidation. The microbiological activity continues even after refrigeration and packaging, though at a reduced rate.

The factors that make fresh meat unsaleble are changes in colour, odour, taste and texture. The pigments present in fresh meat are proteins like hemoglobin and myoglobin. Hemoglobin does the function of transfer of oxygen from the blood and myoglobin acts as a storage mechanism of oxygen in cells. Myoglobin has a purple red colour, which is the characteristic colour of fresh meat when it is first cut. In presence of oxygen, there is formation of oxymyoglobin, which imparts a bright red colour to the meat. In the absence of oxygen, oxymyoglobin gets reconverted to myoglobin.

An undesirable brown colour is formed due to metamyoglobin when meat gets exposed to air for a few days. Yet another form of discolouration on the surface of the meat, which is dark reddish-brown colour is due to loss of moisture. Dehydration of meat on the surface results in concentration of the pigments. Further concentration of pigments occurs, when the interior moisture containing dissolved pigments migrates to the surface and evaporates.

Another factor, which accelerates desiccation and oxidation of meat, is ultraviolet light. Undesirable flavours, odours and textures can occur due to the action of enzymes, molds, bacteria and oxygen if they are not properly controlled. During the storage of fresh meat, the flavour/odour may get affected due to the pick-up of foreign odours or as a result of oxidative rancidity. Hence, the principal role of fresh meat package is:

- To prevent moisture loss.
- To offer the product to the consumers in most desirable colour-red bloom.
- To prevent further bacterial contamination of meat.
- To arrest pick up of foreign flavour and odour by meat.
- To control oxygen transfer.

To prevent dehydration, a relative humidity of 85% to 95% is required during storage. This can be achieved by use of a packaging material, which has a good water vapour barrier. The material should also prevent absorption of odours and flavors from external sources. The control of oxygen permeation requires a compromise between development of ideal colour and prevention of oxidative rancidity of fats. Therefore, it is recommended to use a plastic film with moderate oxygen barrier properties. In order to retain the desired red colour of the meat, a packaging material with good oxygen permeability of 5000ml $O_2/m^2/24$ hrs/10 atmosphere at 24 °C with 100% RH inside the pack and 52% RH outside the pack is required.

Temperature has a very strong influence on the spoilage of meat due to bacteria or molds. Proper storage conditions and selection of good packaging film can help accomplish these objectives. Fresh meat should be stored at 0 °C and 85 to 90% RH.

Packaging Materials and Techniques

Tray with Over-wrap

Retail cuts of fresh meat are generally placed in rigid trays of expanded polystyrene or clear plastic trays, over-wrapped with transparent plastic films. The advantages of using these trays are that they are non-absorbent and aesthetically appealing. Usually blotters are kept at the base of the tray to absorb meat juice. The over wrapping film should have excellent optical properties. Cellophane coated with nitrocellulose on one side has been in use for wrapping fresh meat for a long time. The un-coated side is kept in contact with the meat. Moisture saturation on the inner side of the film increases its oxygen permeability, while nitrocellulose coating on the outer side prevents excessive moisture loss to the atmosphere. Another grade of cellophane with polyethylene coating on one side is used to pack irregular shaped meat cuts.

Plastic Trays with Peelable Lids for Ready-to-Cook Poultry Products.

LDPE may also be used for fresh meat packaging. At gauges lower than 0.001 in, it is sufficiently permeable to oxygen and provides a suitable moisture vapour barrier. One problem, however, has been the condensation of moisture on the inner surface of the polyethylene.

In order to avoid this, LDPE film with minute holes is used. By modifying LDPE with

Vinyl acetate a good clear film can be made, which has suitable oxygen permeability. Polyvinyl chloride is also used for fresh meat packaging. It accounts for an estimated 95% of all applications. The film is highly plasticised in order to provide oxygen permeation and contour wrapping. Biaxially oriented polystyrene film is also used to pack some fresh meat packages. Its clarity and permeability provides an aesthetically pleasing and functional package, but it is not durable and has sealing problems.

Meat thus wrapped, can be kept for approximately 10 days at a temperature of 0 °C before it becomes microbiologically unacceptable. However, it would be un-saleable in less than half this time because although still edible, it changes colour from red to an unattractive brown. If during storage, there are fluctuations in temperature up to 5 °C (which is quite likely commercially) the actual shelf-life would be only 2 days. The storage life of meat is dependent on the initial level of bacterial contamination and the temperature during storage. Generally, therefore pre-packed fresh meat are refrigerated as near as possible to −1 °C, which is the lowest temperature at which meat can be stored without freezing it.

Shrink Packaging

Plastic Shrink films are used for wrapping large and uneven cuts of fresh meat. It is a technique in which heat shrinkable polymer film is shrunk around the meat product by application of heat to achieve a skin-tight and compact pack. The packaging film should have structural strength. It should be a good water vapour barrier and be capable of withstanding storage temperature of about – 45 °C.

The advantages of plastic shrink film include neat appearance, ease in handling and a contour fit. Hot tunnels are used to affect a tight wrap. Heat shrinkable polyvinyl chloride, polypropylene, irradiated polyethylene PVDC and rubber hydrochloride are used to shrink wrap fresh meat.

Vacuum Packaging

This technique is used for packaging of primal and sub-primal cuts of buffalo meat. In vacuum packaging, the product is filled in a bag/pouch, air is withdrawn either by nozzle vacuuming or by chamber vacuuming and the bag/pouch is heat sealed thus storing the product in an air free environment. The plastic film used for vacuum packaging must have a high resistance to gases and water vapour with perfect seals and good mechanical strength. Since vacuum packaging provides a barrier to the product from oxygen and moisture, it is suitable for a period of 3 weeks. The most commonly used film for fresh meat vacuum packaging is PVDC (poly vinylidene chloride). It offers low oxygen permeability and shrink characteristics so that large cuts can be kept for up to 21 days with minimum loss of moisture. Typical packaging materials used are:

- Polyethylene/co-polymer coated cellulose film laminates.

- Polyester/polyethylene film laminates.
- Polyamide/polyethylene laminates.

Plastic Pouches for Vacuum Packed Poultry Products.

These are usually used as sachets with PE inside to promote efficient heat seals:

- Laminates of plastic film with aluminium foil.
- PVDC co-polymer films.
- EVA/Saran/EVA laminates.
- Nylon/EVA/laminates.
- PVDC/Polyester/PVDC/PE laminates.
- LDPE/BA/nylon/BA/LDPE.

Vacuum packaging is more a means of keeping meat at better level of quality than the means of increasing shelf-life because it is virtually impossible to remove all the oxygen (small quantities remain trapped within the product). Also, some anaerobic organisms, which are not affected by oxygen, are not reduced.

A shelf-life of up to three weeks for fresh meat is achieved by vacuum packaging, but it is not popular at retail level because of purple red colour of meat. During the period of three weeks, the enzymes in the meat continue the process of tenderising. At the end of the storage period the meat blooms. Vacuum packaging ensures a life of 8-10 weeks at 0 °C for buffalo meat. Lamb and pork have a short life as compared to beef. Lamb has higher pH while pork has higher initial load of bacteria. However, vacuum packaged boneless joints are reported to have a shelf-life of 2 weeks at 1 °C. Vacuum packaging in cryovac barrier bags is also used for storage and transport of fresh meat products.

Skin Packaging

Another development that offers advantages for presentation as well as packaging design variety is skin packaging. The process allows the packaging film to conform exactly to the profile of the product. This gives rise to many opportunities for enhanced product presentation as well as further improving the integrity of the pack itself. In a skin pack, the product becomes the die for the thermoform packaging operation. The semi-rigid bottom web may or may not be thermoformed. The top web is heated in an evacuating chamber until it is near its melting point, at which it drapes over the product and forms a skin around all the contours. Upon sealing and cooking, it retains its new shape, ensuring intimate contact with the product, irrespective of surface irregularities. Skin packs are prepared with an oxygen barrier plastic film.

Modified Atmosphere Packaging

The use of MAP technology in meat industry has shown increasing trend during the past decade. MAP extends the shelf-life of meat and meat products under refrigerated storage and maintains, colour, texture and flavour of the product for a considerably longer time in flexible plastic films. Application of MAP improves shelf-life of raw chilled meat with reference to colour, delaying bacterial spoilage and minimising exudate loss.

Meat requires different conditions for packaging, beef needs high oxygen content to maintain a bright brick-red colour. Chicken needs little oxygen and reduced amount of carbon dioxide. Pork with its high fat content needs less oxygen and some nitrogen for glow.

MAP involves placing fresh meat in an environment in which the oxygen availability is altered from that in the air. This is commonly done either through the removal of air/oxygen by vacuum and then back flushing with carbon dioxide, nitrogen or a combination of the two. Carbon dioxide inhibits the growth of typical aerobic meat spoilage microorganisms. Carbon dioxide also inhibits mold growth, but the same is true for complete nitrogen flush and vacuum packaging in which the oxygen is almost totally removed. In all three processes, low oxygen availability atmosphere are created, inhibiting aerobic meat spoilage organisms. It has been reported that, in general, the use of 20% carbon dioxide and 80% oxygen is a good proportion with respect to balancing the attractive meat colour and retarding the bacterial growth thus extending the shelf-life. A mixture of 70% oxygen, 20% carbon dioxide and 10% nitrogen at 0 °C slows bacterial growth so significantly that microbially caused discolouration, sliminess and aroma changes are eliminated. A mixture of 80% carbon dioxide and 20% nitrogen is better bacteriostatically, although some colour loss occurs. Fresh pork in 10% carbon dioxide can be stored for a period of 40 days at 4 °C.

Table: Typical gas mixtures used for some of the meat products.

Product	Temperature in °C	O_2%	CO_2%	N_2%
Fresh Meat	0 – 2	70	20	10
Cured Meat	1 – 2	0	30	70
Pork	0 – 2	80	20	0
Poultry	0 – 2	0	20 - 40	80 - 60

Similarly, the blends of gases containing 90% carbon dioxide and 10% nitrogen; 85% carbon dioxide and 15% nitrogen are equally effective in providing packages with good colour retention, good bacterial check and good product and package appearance. The typical gas mixtures used for some of the meat products are given in above table. Changes in meat, fish and poultry as brought about by modified atmosphere are given in table.

Table: Changes in meat, fish and poultry as brought about by modified atmosphere.

Changes	Results
Enzymatic ageing process	Unaffected.
Microbial spoilage	Increased CO_2 reduces growth of aerobic spoilage.
	Organisms by penetrating membranes and lowering Intracellular ph.
Fat oxidation	Reduced O_2 reduces oxidation of fats, although oxidation can still occur at low O_2 tensions.
Oxidation of myoglobin	Increased CO_2 promotes metamyoglobin formation and colour darkening.

For these types of packages, the packaging film should have good gas barrier and sealability properties and have optimum clarity and antifogging properties so that packs maintain their clear appeal in or out of the chill display. For most MAP applications, conventional multilayer, high barrier films are used. The various types of packaging films used for MAP of different meat products are given in table.

Table: Plastic packaging films used for ma packed meat products.

Types of Film/Packaging Structure	Product
Curlon 863 (Nylon/Curpolymer-surlyn)	Beef rib eye rolls of 1" thick
Nylon – Polyester Vacuum bags B 540, Cryovac, W.R. Grace	Ground beef
Saran coated PVC bags	Beef strip loins
PA80/PE 100-PE 100/PA80/PE 100	Smoked pork loin
Mylothene 12/80/5	Sausages
PETP/X/PE/X, 12/5/70/5 laminated plastic bags	Fresh pork chops
Thermoformed plastic pouches (lower part PVC/PE, 250/70 and cover PETP/X/PE, 12/5/70)	Processed ham
Cryovac barrier bags	Ground beef patties
Foil laminate pouches	Boneless pork loin

To attain shelf-life extension in packaging and distribution of fresh meat in MAP, it is very important that temperature be maintained as near to 0 °C as possible. As compared to vacuum packs, gas packed products have a better appearance but have a comparatively shorter life. These packages also require more space during storage and use more packaging material. In vacuum packs, since the drip loss is negligible, higher shelf-life can be attained. The disadvantages of vacuum packs are the unattractive purple colour of the meat and the problems of packaging meat cuts with bones, which can pierce through the film.

Packaging of Frozen Meat

Preservation of meat by freezing, offers the greatest advantages of increase in shelf-life, inhibition of bacterial growth and preservation of fresh texture and flavour. If frozen meat is not properly packed there is continuous dehydration from the surface resulting in freezer burn. This condition affects the surface texture and colour. Meat fat is also prone to the development of oxidative rancidity if a good oxygen barrier is not used. Hence, frozen meat needs protection against:

- Dehydration and loss of surface texture.
- Moisture loss.
- Temperature fluctuations.
- Rancidification.
- Pick up of odours/flavours.
- Expansion and contractions which occur during freezing and thawing.

Packaging Requirements

A suitable packaging material must, therefore, have a very low moisture vapour and oxygen transmission rate. The material should also be durable at freezer temperature, have very high wet strength and be impermeable to odour and flavours.

Packaging Materials

Plastics are amongst the most commonly used materials for packaging of frozen meat Generally, low density polyethylene (150 - 200 gauge) is used for this purpose as it provides adequate clarity and is stable at low temperatures and is available at low cost. Polyester or Nylon/PE laminates and heat shrinkable low density polyethylene and PVC/PVDC co-polymer films also provide functional properties, besides giving neat appearance to the frozen meat cuts. Shrink packaging also allows convenient handling of the product.

Frozen meat may be in the form of chunks, minced or various cuts. The unit packs consist of 1,2,4 or 8 lbs in LDPE bags of 250 - 350 gauge. After placing the meat in the bags, the bags is folded and then the packed product is blast frozen at – 40 °C for a period of 4 to 12 hrs depending upon the size and shape of the package. After freezing, unit packages are packed and stored in corrugated boxes made of either paper or plastic, which are either waxed internally, or on both the surfaces. These boxes are stored at -20 °C and the expected shelf-life for the product is around one year.

Packaging of Cured Meat

Cured meat like ham, bacon, luncheon meat and frankfurters have a shelf-life of 12-15 days at 4 °C depending upon the degree of curing. The attractive pink colour present in cured meat is due to a pigment called nitrosomyoglobin. Although this enzyme is more stable than oxymyoglobin, it is readily oxidised to metamyglobin. The colour of the meat can fade due to the action of light.

Packaging Requirements

The role of package for cured meat is to minimise light fading by preventing the entry of oxygen and loss of moisture. The spoilage due to growth of bacteria and yeast is expected to be reduced by freezing or refrigeration. The package must be able to withstand low storage temperatures and also present an attractive appearance. The packaging material should be a good oxygen and water vapour barrier. It should be flexible enough to make a close surface contact with meat. The packaging film should be capable of lamination or co-extrusion and hermetical sealing.

Meat and Poultry Products Packed in Laminated Plastic Pouches.

Packaging Materials

- Over wrapping: Short-term storage of cured meats can be done by over wrapping the product with polyethylene. Some other films like Poly Vinyl Chloride (PVC) Polyvinylidene Chloride (PVDC) provide tight fitting over-wraps. Aluminium foil/paper laminate can protect the cured meats against light. Another type of package called the chub pack is used for some ground products. Chub packs are tubes stuffed with soft products which is twist tied or clipped at each end. The packaging film used for chub packs is generally polyethylene.
- Shrink Packaging: Ham and other large irregular cuts of meats are packed in

shrinkable PVDC/PVC co-polymer films where the air is evacuated and the contour is over wrapped either by immersing in hot water (90 – 95 °C) or passing through hot air tunnel to effect the shrink.

- Vacuum Packaging: For long term storage of bacon blocks, luncheon meat etc. vacuum packaging is the right choice. This technique assures a shelf-life of 5 months at 4 °C and 6 - 8 months at –18 °C. The following laminates are used for vacuum packaging:
    - Cellophane/PVDC/LDPE.
    - Polyester/PVDC/LDPE.
    - Polyamide/PVDC/LDPE.
    - Metallised Polyamide/Ethyl vinyl acetate (EVA).
    - EVA / PVDC/EVA.
    - Polyamide/LDPE/ionomer.
    - PVC / PVDC/LDPE are used for packing sliced luncheon meat.

In all the above cases, plastics provide a wide range of packaging solutions.

Packaging of Thermo-processed/Cooked Meat

Most cooked meat is canned and has a long shelf-life of over two years. Thermal processing is usually done above 100 °C by applying pressure. Generally, hermetically sealed rectangular tinplate containers with easy open devices are used. Meat products like patties, sausages, nuggets and meat balls are packaged in pouches made of polyethylene, polypropylene, PVDC, rubber hydrochloride etc. for short term storage testing for 10-12 days at 4 °C.

Products like corned beef, corned pork, meat gravies, meat soups, liver sausages, chicken curry, boneless chicken etc. are hermetically sealed and cooked to make commercially sterile for long term storage at room temperature. Metal containers are used for this purpose. Canned meat products are shelf-stable for a number of years at room temperatures. Tinplate is mostly used to pack these products. They are coated on the inner side with sulphur - resistant lacquer. Shallow drawn aluminium cans with internal lacquer are also used to pack some products.

Packaging of Dehydrated Meat

Dehydration is a successful means of preserving many meat products with proper packaging because they are highly susceptible to oxidation resulting in rancid odour. The packaging materials used are tinplate cans. Metal foil/plastic film laminates are used to pack compressed bars of dehydrated minced meat with inner cellophane and outer paper/aluminum foil/PE laminate. This pack is said to be shelf-stable for one year. Flexible pouches suitable for vacuum and modified atmosphere packages made from polyester/PE/Aluminium foil/PE or cellophane/PE/Aluminium foil/PE laminates are also used.

Packaging of Meat Products

Various meat products like sausages, frankfurters, luncheon meat, corned beef, cooked ham, bacon, meat pies etc., are made from comminute and seasoned meat blended with various cereals, binders, spices and preservatives. Their packaging requirements are similar to fresh meat such as trays and films, shrink film bags, envelopes and pouches.

Pouch-in-Carton for Ready-to-Cook Poultry Products.

Some varieties of smoked products such as sausages, salami, frankfurters are vacuum packed in pouches with or without trays or card board inserts. Sliced meat products like ham, bacon are loosely packed in a pouch or they are singled or stacked and vacuum packed. Some products like chicken leg masala, chicken-in-curry, meat-in-brine etc. are packed in metal cans. The laminates used for vacuum packaging are combinations of cellophane/PVDC/polyethylene or polyester/PVDC/polyethylene. Some plastic films can be drawn into cavities, which make neat package for slices. These packages are made from combinations of polyethylene/PVDC/polyamide or polyester.

Packaging of Poultry

Poultry meat is high in protein, low in calories and easy to chew and digest, but poultry fat is unsaturated and is very prone to the development of oxidative rancidity. Shelf-life of poultry varies according to type of processing, nature of processing environment, initial flora and post-slaughter treatment. Packaging of poultry meat is undertaken immediately after the dressing operations are over. Unpacked refrigerated storage may result in surface dehydration, freezer burn, characterised by surface dis-colouration, tough texture and diminished juiciness as well as flavour loss. Poultry is usually packed as whole dressed poultry, cut up poultry and poultry organs. Dressed poultry have a shelf-life of 5-7 days at refrigerated storage conditions of 0-5 °C.

Chilled and Frozen Poultry Products in Flexible Plastic Pouches and Thermoformed Plastic Tray.

Packaging Materials

- Over-wraps: Packaging of whole dressed chicken halves or cut-up parts are done in plastic films like polyethylene, polypropylene, PVDC, rubber hydrochloride or nylon-6. These are films of 150 to 200 gauge in thickness. Polyethylene is the most widely used packaging material in our country because of low cost and easy availability. These thermoplastic film sheets can be fabricated into bags. Each dressed eviscerated bird is inserted into a bag. Giblet of individual bird is wrapped in waxed paper or parchment paper and placed into the body cavity before bagging. The problem of body fluid accumulation is avoided by putting an absorbent pad or blotter on the pack of each bird to soak up the liquid. The bag is then heat sealed or twist-tied or clipped shut.
- Tray with Over-wraps: Small whole dressed chicken, broilers, roasting chickens are placed in a polystyrene foam tray and over-wrapped with transparent plastic film. A blotter underneath absorbs the excessive meat juice accumulated. Chicken thus wrapped has a shelf-life of 7 days at 4 °C in refrigerator.
- Shrink Film Over-wraps: Many thermoplastic films such as polyethylene, polypropylene, poly vinylidene can be biaxially oriented to stay stretched at ambient temperature. Dressed chicken is overwrapped with such films and passed through hot air tunnel or dipped in water tub maintained at 90 °C for a few seconds to effect shrinkage of the film.
- Vacuum Packaging: The ideal materials for vacuum packaging of poultry are laminates of polyester/polyethylene (PE), polyamide/polyethylene, PVDC co-polymer film, nylon/EVA.
- Modified Atmosphere Packaging: In this technique, the atmosphere surrounding the product in the pack is modified by flushing carbon dioxide, nitrogen and oxygen alone or in combination. Mixture of 60% nitrogen and 40% carbon dioxide or 50% nitrogen and 50% carbon dioxide is ideal for modified atmosphere packaging of chicken meat.

Dressed Chicken Packed in Plastic Pouches.

Modern Trends in Meat and Poultry Packaging

With the development of new meat products entering the market, high quality alternative packaging materials are also emerging. The consumer is becoming more discerning in his or her choice of food products and there is a trend, which shows a shift from traditional food items and eating habits. Some of the recent trends in the meat packaging industry are covered here.

Retortable Flexible Pouch

A retort pouch can be defined as a flexible package into which a food product is placed, sealed and then sterilized at temperatures between 110 °C to 140 °C. The finished product is commercially sterile, shelf-stable and does not require refrigeration. A retortable pouch is made of a laminate of three layers held together by adhesive. The outer layer made of polyester, polyamide or oriented polypropylene provides support and physical strength to the composite.

The middle layer of aluminium foil acts as a barrier against water vapour, gases and light. The inner layer of polyethylene, polypropylene or PVC provides heat sealability and food contact. The different laminates used for a retort pouch are polyester/aluminium foil/modified high-density polyethylene or polyester/aluminium foil/polypropylene-ethylene co-polymer. A retortable pouch like the metal can, can be sterilised by heat and it has the advantage of lower cooking time as it has a thinner profile than metal cans. They are not required to be stored at refrigerator temperatures and are shelf-stable, as canned food. Also, these types of packages require less storage space and are lighter in weight as well.

Roast-in-Bags

It is an oven stable vacuum skin package that can be used to cook meat at a temperature up to 204 °C. It is fabricated from polyethylene terephthalate (PET) film due to its unusual properties such as it does not become brittle with age, has long shelf-life, has resistance to most chemicals and moisture and is dimensionally stable.

Microwave Packages

Convenience food falls into two categories, frozen and retortable. The current trend in frozen food is dual oven-ability i.e. products that can be heated in a microwave oven and conventional oven. Shelf-stable retortable food is better suited to microwave heating. Owing to the growing importance of microwave ovens, other materials are overtaking the conventional aluminium trays.

While selecting thermoplastics for dual ovenable packages, the critical properties to be considered are dimensional stability up to 200 °C to 250 °C, good impact strength at freezer temperatures and microwavability. Heat resistant plastic trays made from materials like polyester, polypropylene, nylon and polycarbonate can be used in combination or as monolayers. These trays are closed with heat-sealable lidding materials or over-wrapped or shrink wrapped or sealed inside a microwave bag. Thus, due to their flexibility of forming different shapes and sizes on in-line formation, rigid plastic trays dominate the microwavable packaging.

Polyester coated paperboard cartons are also used as microwave packages. These containers can be formed on a conventional tray making, carton forming and folding carton making machine. Crystallized polyester containers are also very popular for microwave packaging as well as for conventional oven cooking. These trays are very stiff and can be sealed in a high-speed tray sealing machine with transparent or non-transparent lidding material. They are easy to handle, sturdy, attractive, cost competent and can be compartmentalised for multicomponent food items. They are self-serving and reusable.

Cryovac Packages

Pre-cooked meat packages with shrink film are an innovation with equipment made and supplied by Cryovac. In this system, there is a double chamber machine. In the longer chamber the meat is placed in the film bag and the mouth of the bag is drawn into the smaller chamber. In the longer chamber a low level of vacuum is just drawn to balloon the bag in order to eliminate formation of air pockets.

A high level of vacuum is then drawn into the smaller chamber where the mouth of the bag is placed. When both the chambers are at maximum vacuum, an automatic clipper closes the bag. The packaged meat is then sent to the shrink tunnel. Cryovac package forms a skin tight package. This together with the use of a film with low oxygen permeability stabilises the product and increases the shelf-life.

Packaging of Food Grains and Flours

Cereals are considered to be fruits of cultivated grasses. They are the members of the monocotyledonous family Gramineae. The principal cereal crops are wheat, barley, oats, rye, rice, maize, sorghum and the millets. Cereals have been important crops for thousands of years and the successful production, storage and use of cereals has contributed largely to the development of modern civilization. Today cereals and cereal based products are an important part of the diet in most countries, and each year new products based on cereals are developed and marketed to increasingly urbane consumers.

Grains

The cereals are harvested, transported and stored in the form of grains. The anatomical structure of all cereal grains is basically similar, differing from one cereal to another in detail only. The mature grain of the common cereals consists of carbohydrates, proteins, lipids, minerals and water, together with small quantities of vitamins, enzymes and other substances, some of which are important nutrients in the human diet. Carbohydrates are quantitatively the most important constituents, forming 77 to 87% of the total dry matter.

The lipids in milled cereals are likely to undergo two types of deterioration: Hydrolysis from endogenous lipases, and oxidation from endogenous lipoxygenases or molecular oxygen. Destruction of the grain and the fragmentation that occurs in milling promote deterioration by bringing the lipid and the enzyme together. The structure and quality of the grains changes due to various processing steps applied. The threats to grain during storage are moisture, temperature, fungi, bacteria, insects and other pests. If the grain moisture content can be controlled, the threats due to temperature-rise, fungi and insects can be largely controlled.

It is not the moisture content that is important for long term storage but rather the water activity (a_w), a safe moisture content for long term storage of grain and oil-seeds usually being accepted as one in equilibrium with 70% relative humidity, i.e. 0.70 a_w. Above 0.75 a_w molds will grow rapidly during storage, with subsequent deterioration. Thus, the loss of product. Moisture sorption isotherms for wheat, barley and maize and all have a typical sigmoid shape. The moisture content equivalent to 0.70 a_w varies quite significantly among the different grains, thus stressing the utility of expressing stability in terms of aw rather than moisture content.

Storage of grains take many forms, ranging from piles of unprotected grains on the floor, underground pits or containers, and piles of sacked grain, to storage bins of many sizes and shapes. It is essential that the grain has been dried to a moisture content equivalent to 0.70 a_w or less before packaging and storage. Consumer packages for grain commonly consist of heat sealed pouches made from LDPE; these provide a satisfactory moisture barrier and result in the required shelf lives for the grains.

Wheat

The wheat grain is a live, respiring organism commonly carrying endemic fungi. Respiration is slow at 14% moisture content and 20 °C, but increases as moisture content and temperature increase. The process of respiration generates heat, which is difficult to remove since wheat is a poor conductor as well as CO_2 and water vapour, resulting in loss of weight. Except the grain is turned over to allow evaporation of the water, it will sweat and become caked in the storage.

Moisture contents of 16% to 30% can support fungal growth in wheat. Thus, there is the risk of mycotoxin production. Above 30% moisture content, it is susceptible to bacterial infestation, leading to spoilage, excessive heat production and possibly charring. Insect life also becomes more active as the temperature rises and, because of their respiration, live insects in grain also increase the grain temperature. Deterioration during storage is intensified by mechanical damage during harvesting since microorganisms attack damaged grains more readily than intact grains. A drop of 3% in moisture increases the shelf life four times. The major problems at these higher moisture contents include accelerated wheat enzyme activities and microbial spoilage.

Rice

Rice is consumed in large quantities in many countries. Brown or unmilled rice is more nutritious than milled rice, but storage stability problems and a traditional consumer preference for whole rice have limited the quantities of brown rice packaged and sold for direct consumption. A major limiting factor to less use of brown rice is accumulation of free fatty acids in rice stored under warm and humid conditions. Fatty acids can be released by lipase activity present in the rice bran layer of damaged grains and by high lipase-containing microorganisms adhering to rice.

In a study to evaluate the effects of CO_2 gas flushing on the shelf life of brown rice, greater stability of the rice was obtained at refrigerated (4 °C) storage when it was packaged in nylon-EVA copolymer laminate film bag rather than a regular plastic bag. However, no differences were found when stored at 24 °C. The gas flushed bags formed a hard, rigid pack similar to a vacuum packaged product and maintained this appearance for at least 3 years. It was suggested that if bulk rice in warehouses could be stored as brown rice instead of as rough rice in laminate or gas flushed bags, savings of at least 20% by weight and 30 to 35% by volume should result.

Flour

Wheat flour is the ground or milled form of wheat grain in which the bran and germ are partly removed and the remainder is comminuted to a suitable degree of fineness. It has been suggested that for long periods of conservation, flour should be stored in a closed atmosphere. Under these conditions, flour acidity increases due to the accumulation of linoleic and linolenic acids. Flour is stored commercially in bags or in bulk bins. The threats to flour in storage are similar to those of wheat in storage, viz., mold and bacterial attack, insect infestation, oxidative rancidity and eventual deterioration of baking quality. The expected shelf life of plain white flour packaged in paper bags and stored in cool and dry conditions and protected from infestation is 23 years. The rate of increase in acidity increases with storage temperature and decreasing flour grade by increasing ash residue in the flour. Thus, the shelf life of brown and wholemeal flours is shorter than that of white flour. Freedom from insect infestation during storage can be guaranteed only if the flour is free from insects at the time of packing, and even the storage area is free from insect infestation. The use of multi-walled kraft paper bags, sometimes with an LDPE liner, completely eliminates any infestation hazard.

The optimum moisture content for the storage of flour is related to the intended shelf life, the barrier properties of the packaging material, and the ambient conditions. For use within a few weeks, flour can be packaged at 14% moisture content, but at moisture contents higher than 13%, mustiness due to mold growth may develop over time. At moisture contents lower than 12%, the risk of lipid oxidation and the development of rancidity increases. Australian millers maintain flour moisture contents below 12% for shipment to warm and humid areas, while in the United Kingdom wheat flour is stored at maximum moisture content of 12%.

Reference to a moisture sorption isotherm shows that a moisture content of 12% corresponds to a_w of 0.50. As with wheat and other grains, the moisture content is an unreliable means to measure stability and what is required are moisture sorption isotherms at various temperatures so that the moisture content which corresponds to the maximum a_w for stability at a particular storage temperature can be determined. This will vary depending on the type of grain as well as the variety. But, data on the shelf life of flour at various a_w's and values for critical moisture contents are missing, and it is therefore difficult to specify precisely the type of moisture vapour barrier required in a package.

Notwithstanding the lack of information about critical moisture contents and shelf lives as a function of a_w, bags made from cotton twill or paper have been used successfully for decades for consumer packs of flour. Kraft paper bags with an LDPE liner provides additional protection, thus, a longer shelf life.

Breakfast Cereals

Breakfast cereals can be categorised according to (1) the amount of cooking required, (2) the form of the product, and (3) the cereal used as raw material. All cereals contains large amount of starch which in its natural form is insoluble, tasteless and unsuitable for human consumption. It must be cooked or processed to make it digestible and acceptable. In hot cereals the cooking is carried out in the home, while ready-to-eat cereals are cooked during manufacture or processing.

If the cereal is cooked with excess water and moderate heat like in boiling, the starch gelatinizes and becomes susceptible to starch-hydrolysing enzymes in the human digestive system. If the cereal is cooked with a minimum of water or without water but at higher temperatures as in toasting, nonenzymatic browning between protein and reducing sugars may occur and there may be some depolymerisation of the starch.

Deterioration in Breakfast Cereals

There are five major modes of deterioration to be considered while selecting packaging materials for breakfast cereals. They are (1) Moisture gain resulting in loss of crispness, (2) Lipid oxidation resulting in rancidity and off-flavours, (3) Loss of vitamins, (4) Breakage, resulting in an aesthetically undesirable product and (5) Loss of aroma from flavoured product.

The shelf life of breakfast cereals depends on the content and quality of the oil contained in them. Thus, products made from cereals having a low oil content such as wheat, barley, rice and maize grits (oil content: 1.5 to 2.0%) have a longer shelf life than products made from oats (oil content: 4 to 11%, average 7%). Although whole corn has a high oil content (4.4%), most of the oil is contained in the germ which is removed in making grits.

Packaging of Breakfast Cereals

Packaging of breakfast cereals has traditionally been in fibreboard boxes with a super calendered waxed glassine liner. Now the glassine liner has been replaced by various plastic materials, in particular with thin gauge HDPE which is usually folded rather than heat sealed. HDPE coextruded with a thin layer of EVA copolymer is a recent introduction, the EVA copolymer allowing a lower heat seal temperature and offering the consumer an attractive and peelable seal.

When the cereal product is not hygroscopic and holds a satisfactory texture when in equilibrium with the ambient conditions, a liner may not be needed for moisture protection and may even serve to entrap rancid aromas. In such case, either no liner or one which is vapour permeable may be used. As well as providing a barrier to vapour, the liner must also restrain cereal aromas within and at the same time prevent foreign odours entering the packaged product. It should also be reclosable to protect the remaining cereal in the package.

To minimize oxidative rancidity in the cereals it is important that the package eliminate light. Excluding oxygen may be of limited assistance in extending shelf life. For this reason, most processors do not bother to use good oxygen barrier packaging. In a study of the storage stability of flaked oat cereal packaged in materials of different oxygen barrier properties with and without the addition of an iron-based oxygen absorber, the absorber delayed lipid oxidation provided that it was used with a packaging material such as PVC/PVDC copolymer-coated PP-LDPE that was a good oxygen barrier. The use of an anti-oxidant in the package liner has been shown to be successful in extending shelf life but is not generally permitted in most countries.

In shredded cereals, rancid odours accumulates if shredded wheat is stored in an air-tight container, due to the higher unsaturated fat content. Therefore, it is usually sold in breather boxes without linings. When so packaged, the product is just as stable as any other prepared cereal, except that moisture absorption takes place much more readily.

The vitamin and mineral fortification of cereals is widely used in many countries. The major cause of vitamin loss in packaged cereals is the temperature of storage. In a study on the effects of processing and storage on micronutrients in breakfast cereals, it was found that micronutrient loss is not a major factor in determining the shelf life of dry cereals. There was no considerable loss of added vitamins during normal shelf lives, with the possible exception of vitamin A and C. Vitamin A survived six months at room temperature with no measurable loss.

The rigidity and compression resistance of the carton together must provide the necessary resistance to product breakage throughout production line operations, storage and distribution. Rigidity also prevents the bulging of the carton. Protecting breakfast cereals from breakage does not appear to be a problem using currently available carton stock and carton designs.

Loss of flavour is a problem with certain cereal products to which fruit flavours have been added before packaging. In these cases, loss of flavour results in the product being considered to be at the end of its shelf life by the consumer. A study evaluating HDPE and glassine as liner found that the permeability coefficients of d-limonene (common flavour component in citrus products) in the HDPE liner were three to four times higher than in glassine. It was also found that the solubility of d-limonene in the glassine liner was significantly lower than in the HDPE liner for the same vapour pressures. Hence, equilibrium distribution of the limonene vapour between a product such as a fruit flavoured cereal and the respective liners will result in a lower limonene concentration within the glassine liner, and scalping of the flavour can be expected to be more significant in the HDPE liner.

Permissions

All chapters in this book are published with permission under the Creative Commons Attribution Share Alike License or equivalent. Every chapter published in this book has been scrutinized by our experts. Their significance has been extensively debated. The topics covered herein carry significant information for a comprehensive understanding. They may even be implemented as practical applications or may be referred to as a beginning point for further studies.

We would like to thank the editorial team for lending their expertise to make the book truly unique. They have played a crucial role in the development of this book. Without their invaluable contributions this book wouldn't have been possible. They have made vital efforts to compile up to date information on the varied aspects of this subject to make this book a valuable addition to the collection of many professionals and students.

This book was conceptualized with the vision of imparting up-to-date and integrated information in this field. To ensure the same, a matchless editorial board was set up. Every individual on the board went through rigorous rounds of assessment to prove their worth. After which they invested a large part of their time researching and compiling the most relevant data for our readers.

The editorial board has been involved in producing this book since its inception. They have spent rigorous hours researching and exploring the diverse topics which have resulted in the successful publishing of this book. They have passed on their knowledge of decades through this book. To expedite this challenging task, the publisher supported the team at every step. A small team of assistant editors was also appointed to further simplify the editing procedure and attain best results for the readers.

Apart from the editorial board, the designing team has also invested a significant amount of their time in understanding the subject and creating the most relevant covers. They scrutinized every image to scout for the most suitable representation of the subject and create an appropriate cover for the book.

The publishing team has been an ardent support to the editorial, designing and production team. Their endless efforts to recruit the best for this project, has resulted in the accomplishment of this book. They are a veteran in the field of academics and their pool of knowledge is as vast as their experience in printing. Their expertise and guidance has proved useful at every step. Their uncompromising quality standards have made this book an exceptional effort. Their encouragement from time to time has been an inspiration for everyone.

The publisher and the editorial board hope that this book will prove to be a valuable piece of knowledge for students, practitioners and scholars across the globe.

Index

S

T

V

W

www.ingramcontent.com/pod-product-compliance
Lightning Source LLC
LaVergne TN
LVHW080616210926
848485LV00003B/169